Financial Intermediation and Deregulation

Contributions to Economics

Lars Olof Persson/Ulf Wiberg
Microregional Fragmentation
1995. ISBN 3-7908-0855-5

Ernesto Felli/Furio C. Rosati/
Giovanni Tria (Eds.)
The Service Sector: Productivity and Growth
1995. ISBN 3-7908-0875-X

Giuseppe Munda
Multicriteria Evaluation in Fuzzy Environment
1995. ISBN 3-7908-0892-X

Giovanni Galizzi/
Luciano Venturini (Eds.)
Economics of Innovation: The Case of Food Industry
1996. ISBN 3-7908-0911-X

David T. Johnson
Poverty, Inequality and Social Welfare in Australia
1996. ISBN 3-7908-0942-X

Rongxing Guo
Border-Regional Economics
1996. ISBN 3-7908-0943-8

Oliver Fratzscher
The Political Economy of Trade Integration
1996. ISBN 3-7908-0945-4

Ulrich Landwehr
Industrial Mobility and Public Policy
1996. ISBN 3-7908-0949-7

Arnold Picot/Ekkehard Schlicht (Eds.)
Firms, Markets, and Contracts
1996. Corr. 2nd printing 1997.
ISBN 3-7908-0947-0

Thorsten Wichmann
Agricultural Technical Progress and the Development of a Dual Economy
1997. ISBN 3-7908-0960-8

Ulrich Woitek
Business Cycles
1997. ISBN 3-7908-0997-7

Michael Carlberg
International Economic Growth
1997. ISBN 3-7908-0995-0

Massimo Filippini
Elements of the Swiss Market for Electricity
1997. ISBN 3-7908-0996-9

Giuseppe Gaburro (Ed.)
Ethics and Economics
1997. ISBN 3-7908-0986-1

Frank Hoster/Heinz Welsch/
Christoph Böhringer
CO_2 Abatement and Economic Structural Change in the European Internal Market
1997. ISBN 3-7908-1020-7

Christian M. Hafner
Nonlinear Time Series Analysis with Applications to Foreign Exchange Rate Volatility
1997. ISBN 3-7908-1041-X

Sardar M.N. Islam
Mathematical Economics of Multi-Level Optimisation
1998. ISBN 3-7908-1050-9

Sven-Morten Mentzel
Real Exchange Rate Movements
1998. ISBN 3-7908-1081-9

Lei Delsen/Eelke de Jong (Eds.)
The German and Dutch Economies
1998. ISBN 3-7908-1064-9

Mark Weder
Business Cycle Models with Indeterminacy
1998. ISBN 3-7908-1078-9

Tor Rødseth (Ed.)
Models for Multispecies Management
1998. ISBN 3-7908-1001-0

Michael Carlberg
Intertemporal Macroeconomics
1998. ISBN 3-7908-1096-7

Sabine Spangenberg
The Institutionalised Transformation of the East German Economy
1998. ISBN 3-7908-1103-3

Hagen Bobzin
Indivisibilities
1998. ISBN 3-7908-1123-8

Helmut Wagner (Ed.)
Current Issues in Monetary Economics
1998. ISBN 3-7908-1127-0

continued on page 151

Tobias Miarka

Financial Intermediation and Deregulation

A Critical Analysis of Japanese Bank-Firm Relationships

With 2 Figures and 26 Tables

Physica-Verlag
A Springer-Verlag Company

Series Editors
Werner A. Müller
Martina Bihn

Author
Dr. Tobias Miarka
Berlin Science Center (WZB)
Reichpietschufer 50
D-10785 Berlin
Germany

ISBN 978-3-7908-1307-4 ISBN 978-3-642-52425-7 (eBook)
DOI 10.1007/978-3-642-52425-7

Cataloging-in-Publication Data applied for
Die Deutsche Bibliothek – CIP-Einheitsaufnahme
Miarka Tobias: Financial intermediation and deregulation: a critical analysis of Japanese bank firm relationships; with 26 tables/Tobias Miarka. – Heidelberg; New York: Physica-Verl., 2000
(Contributions to economics)
Zugl.: Berlin, Humboldt-Univ., Diss., 1999
ISBN 978-3-7908-1307-4

Diss., Humboldt-Universität zu Berlin, 1999.

Physica-Verlag is a company in the BertelsmannSpringer publishing group.

Softcover Design: Erich Kirchner, Heidelberg

SPIN 10769177 88/2202-5 4 3 2 1 0 – Printed on acid-free paper

In memory of my father,

Hans-Jürgen Miarka

The author emphasizes the importance of the Kaisha Data Bank which was developed at the Berlin Science Center with the financial support of the Deutsche Forschungsgemeinschaft. Indeed, the Kaisha Data Bank is a pool of data that are most important for new insights into the behavior of Japanese firms on gaining global markets. Miarka's book is an important contribution to the current debate on bank-firm relationships. It is also a highly valuable contribution to resource-oriented industrial economics.

Horst Albach

Acknowledgments

This work would not have been possible without the help and encouragement I have generously received from many people and institutions.

First and foremost, I would like to express my sincere thanks to Horst Albach, who was much more than the advisor for my doctoral thesis. He laid the foundations for this work in many respects. He encouraged me to deepen my interest in the Japanese economy and taught me to identify new and interesting issues in the field of financial economics. By bringing me to the Berlin Science Center (WZB), he provided me with invaluable academic support together with an ideal infrastructure necessary for conducting well-focused doctoral research. I am deeply grateful for the perfect balance between direction and freedom. Certainly, this work would read much differently without his influence on my conceptual framework and economic views.

I am also indebted to Richard Stehle, who did not hesitate to generously take the time to read and discuss my thesis. His early suggestions concerning Japanese capital markets were particularly helpful.

Crucial was the constant feedback and advice of my colleague and friend Michael Tröge. He took his time during countless sessions in his office to provide me with important insights, especially on theoretical issues, and encouraged me to keep on track during the most difficult times, for which I express my kindest thanks.

My empirical work was greatly advanced by the comments and suggestions of Tomaso Duso, Dietmar Harhoff, Christoph Hilbert, Ralph Siebert, Stefan Speckesser, Andreas Stephan and Jiangping Yang.

The privilege to work together with Noriko Fujimura, Ulrike Görtzen, Max-Martin Kahrs, Andreas Moerke, Tomoko Shioda, Masako Sugitatsu, Thomas Westphal and Rita Zobel in the team "Japanese Business Strategies and Global Market Processes" at the WZB, and to co-operate with Tomoki Waragai's team at Waseda University was especially valuable to the discussion of difficult issues concerning the Japanese economy and working with the KAISHA database.

Further colleagues and friends who were of great help during various stages throughout the study are Thomas Brandt, Manfred Dösl, Christian Göseke, Berit Meinhart, Aaron Schlaphoff and Linda Thomas. I owe particular thanks to Manfred Fleischer for his help and encouragement.

My research benefited much from discussions with participants in the WZB conference "Information Processing as a Competitive Advantage of Japanese Firms", the workshop "Evaluation of Firms, Performance and Financial Structure" at Waseda University, the annual meetings of the Association of Japanese Business Studies in 1997 and 1999, and the EEA99 conference, as well as with seminar participants at the Otto Beisheim Graduate School WHU, Humboldt University and Marburg University. I would especially like to thank Kazuo Araki, Hideki Hanaeda, Jaap de Koning, Yoshio Matsumoto, Hiroyuki Okamuro, Manfred Perlitz, Ulrike Schaede, Mark Scher and Adrian Tschoegl. Their comments and suggestions have greatly influenced my work.

I also feel grateful to Herbert Hax and Hirotaka Takeuchi: Participating in the Program of International Management at the University of Cologne and Hitotsubashi University was essential to me, because the first steps of studying Japanese and the Japanese economy were laid there.

My studies and research benefited from many kinds of research support – including grants from the Association of Japanese Business Studies, the Deutsche Bundesbank, the German Academic Exchange Service (DAAD), the German Research Council (DFG) and the Fritz Thyssen Foundation – for which I express my appreciation.

Finally, and most importantly, I would like to thank my family and my partner Katja for their unconditional love and never lasting support. Without their patience and forbearance, I would not have been able to finish this project.

Tobias Miarka

Contents

Chapter 1

Explaining Japanese Bank-Firm Relationships: Introduction and Overview

1.1 Introduction

External pressures for Japan to change have continuously increased as the world has undergone considerable transformations since the Second World War. During the last fifteen years, however, and particularly with the cessation of Cold War hostilities, Japanese society, including its economic system, has undergone in many respects its most dramatic changes since the end of the Second World War.

Within a few decades, Japan has become an economic power with a gross domestic product approaching that of the United States, and has risen to be the world's largest capital exporter. It has developed an economy that established organic ties with other economies throughout the world via trade, direct investment, overseas manufacturing activities, and the international movement of capital. Hence, considering its current position in a world that is moving steadily toward a borderless economy and deepening international interdependence, it is not only of strong interest for other societies to understand how Japan managed to catch up. It is also of great concern for them how it is trying to secure its top position during a time of crisis that has triggered substantial change.

Not long ago, foreigners flocked to Japan to see how things were done. A popular image of Japan was that of a society whose national objective was to catch up with the advanced economies. It was seen as a nation that successfully built its political, governmental, and economic systems to achieve that objective. Japan was especially admired for its far-reaching consensus-based economy, whose general framework promoted the excellent performance of firms in a wide range of industries; this admiration was deserved, if at times exaggerated.

The Japanese economy was studied as a model of efficiency and innovation, and its bank-dominated financial market was envied of being the source of stupendous wealth. Among the most well known concepts and systems that have been widely seen as a prerequisite and safeguard for the economic success of Japanese firms, are the unique relationships between large Japanese firms and their banks. In fact, the image of the overwhelming advantages of Japanese bank-firm relationships has been so deeply rooted in the mind of the broader public and experts alike that even as late as in 1994, 4 years after Japan's economic boom has turned into bust and the banking crisis began to unfold, the discussion was still largely dominated by the advantages of Japanese bank-firm relationships. Despite strong indicators that increasingly hinted at the fact that there is, after all, also a negative side to them, Japanese-style bank-firm relationships were still widely recommended to emerging and transforming economies as a blue-print for economic success.

Things can change. When the scope of Japan's recession and the magnitude of its banking crisis fully surfaced, even long-time observers of the Japanese economy had difficulties developing a clear picture of what had happened and began to wonder about "The Japan Puzzle" (The Economist 1998). Quickly, admiration turned into despair. Within months, experts and advisers who predicted Japan's inevitable and devastating fall rose to dominate the discussion; among them were many who had explained to the world the uniqueness and superiority of "The Japanese Way" just a few years earlier. Too long was the period of daily revelations of new financial catastrophe after "dirty laundry" got a public airing. Too strong were images like Yamaichi's president Shôhei Nozawa in tears when announcing the closure of Yamaichi Securities in November 1997.

By now, confidence in the state apparatus, the economy, and the banks, which used to be the backbone of Japan's financial system, is at a dismal level. An increasing number of Japanese, worried about losing their jobs and their investments in the floundering economy, rather stuff any extra cash they have acquired into their mattresses instead of buying goods and services. This development is only one indicator that reflects the current state of Japan's society. It will be difficult to come even halfway back to the high levels of trust and confidence that used to be implied in Japan's consensus-driven society. To many observers, the once-promoted advantages of the Japanese economic system might not be true after all. Though the most dramatic incidents seem to have passed by now, the fear of a further prolonged recession has not gone away altogether.

1.2 Motivation and Limitations

Keeping abreast for over close to ten years with both Japanese and Western developments in the academic literature, as well as the sentiments of bankers and business persons, we come up with simple questions which may demand complex answers. Which view is right after all? Is it necessary to throw away now the concepts and systems that were put in place to raise Japan to be the second largest economy in the world? What kind of framework will follow to secure Japan's current position?

To find answers to these questions would be much beyond the scope of this work. It would demand the collaboration of experts in a wide range of different fields and disciplines. However, we can contribute to the discussion by focusing on one distinct aspect of Japan's economy, the relationships between firms and banks, which comprises many features of Japan's past success as well as crucial elements of the transformation process today. This focus has many advantages:

Corporate finance, corporate governance, and therefore bank-firm relationships are affected by changes of Japan's environment more directly than politics, public administration or other elements of the Japanese society. Especially with respect to the quickly-changing domestic financial markets that are strongly and most directly exposed to international pressures, banks and individual corporations undoubtedly have to quickly adapt to changing circumstances and have to perpetually transform themselves in order to survive. Among the different institutions of Japan's society, private corporations have in many respects the highest degree of creativity, which is needed to blaze trails in untraversed territory. Hence, aspects like the evolution of bank-firm relationships that are central to the development of corporations are in many respects leading the change that propagates over time, sectors and institutions. We must identify and examine those changes. Furthermore, and most central for the realization of this project, bank-firm relationships are a well-suited testing ground that provides a wide range of reliable and publicly available data necessary for empirical research on a micro level. The qualitative and quantitative firm-level data that we have compiled for our analysis allows us to assess the advantages of panel data analysis. It thus gives us the opportunity to test factors that could not have been observed with most previous data sets.

To summarize: Changes in the Japanese economy, resulting from new and transformed domestic and international pressures, are likely to be reflected more quickly in the reactions of banks and firms than in government actions and data that comprise these reactions are much more readily available. Thus, an empirical investigation of how bank-firm relationships have changed over the past fifteen years may provide one of the best possible insights on the changes in the Japanese economic system as a whole.

1.3 Research Framework

The reader who is looking for a description of how the Japanese economy, its financial markets and its corporate finance are unique and beyond the understanding and tools of analysis of Western-trained economists should rather refer to the writing of the so-called revisionists. Also, you will neither find an emphatic testimony for the structure of Japanese finance as a blueprint for corporate success, nor a pronounced declaration that "the Japanese way of doing things" was condemned to fail after all.

Our aim is rather to provide a perspective with well-grounded economic

Chapter 2

Overview of the Japanese Financial Market

2.1 Introduction

The institutional framework of Japanese financial markets is well known for facilitating close monitoring by commercial banks. In fact, as we have already pointed out in the introduction, firms' heavy reliance on banks to finance corporate expansion is widely perceived as the key to Japanese finance and thus one of the central foundations of Japan's economic success ever since the early 1950s (Corbett 1987; Suzuki, 1987; Ballon and Tomia 1988; Mayer 1990; Hodder and Tschoegl 1993; Aoki and Patrick 1994; among others).

However, during the 1980s and 1990s, the Japanese economy experienced a wide range of transformations. These included substantial reforms of the legal and institutional structure of Japan's financial markets, as well as an exceptional economic boom-bust cycle that triggered Japan's most dramatic banking crisis since the Second World War. Without doubt, all of these developments deeply shook Japan's once mighty banks[1], which used to operate in a most favorable environment where regulatory restrictions provided a scarcity of market alternatives to bank financing. Along with these developments, one can observe a distinct change in the distribution of capital across differently informed sources of finance. This gives rise to a significant and prolonged modification in intermediary lending in general, and the economic role of bank-firm relationships in particular.

Before delving into an analysis of Japanese bank-firm relationships during the 1980s and 1990s, it is therefore necessary to understand the overall framework of the Japanese economy, in which the changes of corporate finance and subsequently the formation of a new position for banks has been taking place. Hence, the aim of this chapter is twofold. First, we want to provide a brief overview of the legal and institutional modifications of Japan's financial markets.

---

1 During the 1980s, the large majority of Japanese Banks had the highest credit ratings and even as late as 1993, seven out of the ten largest banks world wide in terms of equity as well as total assets were still Japanese (The Banker, July 1993: 99). However, in 1999, this number has declined to only two: Bank of Tôkyô-Mitsubishi; Daiichi Kangyô Bank (dpa, August 1999). Credit ratings moreover have declined dramatically. For a chronology of the emergence and resolution of 'problem' banks in Japan between 1991-1997, see Hall (1998), pp. 160-165.

Second, the chapter gives an account of Japan's recent economic development. Both spheres, which interact in a number of crucial ways, heavily influenced Japanese corporate finance, the position of Japanese banks, and consequently the role of Japanese bank-firm relationships.

The rest of the chapter proceeds as follows: In 2.2, we provide a short assessment of the legal and institutional changes that transformed Japan's highly regulated and bank-centered financial market into a much more liberalized one. Chapter 2.3 gives a selected overview of the salient features of Japan's boom-bust cycle. In 2.4 we discuss the changes in corporate finance by large manufacturing firms that have been primarily triggered by institutional, legal and economic changes since the 1980s. Finally, 2.5 closes the chapter with some concluding remarks.

2.2 Legal and Institutional Framework of Japanese Financial Markets

Regulatory constraints play a key role for decisions in Japanese corporate financing. Until the 1980s, the institutional and legal framework of Japanese financial markets was much in favor of monitored debt. However, in the early 1980s, Japan entered an "era of financial liberalization" (Horiuchi 1996: 167). This was at a time when overall economic conditions were rather favorable. Japan's economy had recovered well from the second oil shock at the end of the 1970s, and with the exception of a brief period after the Plaza-Accord in May 1984, the economy had grown at a rapid and relatively stable rate over a period of almost two decades (Itami 1998). In order to secure Japan's competitiveness, the process of financial liberalization and deregulation encompassed a wide range of areas. It made corporate borrowing and raising of equity in both domestic and foreign markets much easier and more common.

2.2.1 Financial Liberalization: The International Level

Concerning Japan's liberalization process on the international level, the most prominent measure is certainly the amendment of the Foreign Exchange and Foreign Trade Control Law (FEFTCL) in December 1980[2]. It substantially liberalized foreign exchange transactions, initiated the internationalization of the

[2] An additional amendment of the Foreign Exchange and Foreign Trade Law, which further opens up the Japanese financial market has been approved in 1997 (MOF 1997).

yen[3], and therefore triggered the internationalization of Japan's financial markets (Kester 1991). Finally gaining access to a wide range of financial sources other than domestic bank loans that are available in international markets, Japanese corporate finance started to move away from reliance on domestic banks and increasingly used other sources of capital, like corporate bonds issued in Europe and later, in the Tôkyô offshore market. Note that large firms listed at one of Japan's main stock exchanges or, even better, in Japan and at least one stock exchange abroad, were particularly well positioned to quickly use the new opportunities of financing. They were already well known and were, in many cases, already familiar with the foreign markets due to the operations of their subsidiaries abroad. As will be discussed below, this development then raised pressure on the domestic market[4]. The amendment of the FEFTCL can thus be considered the central measure for liberalizing the Japanese financial markets on the international as well as domestic level.

In 1984, the process went further ahead at the international level. Among the numerous legal changes were the full-scale liberalization of interest rates and the abolishment of the "real demand principle", which required foreign exchange transactions to be backed by "real" demand for foreign exchange, such as foreign trade. In 1986, the Euromarket was largely deregulated and the Tôkyô offshore market was opened. Liberalization measures continued with, among other changes, the opening of the Euroyen commercial paper market in 1987, the liberalization of residents' overseas deposits in 1989, the relaxation of criteria for resident issues of foreign currency-denominated bonds in 1993, as well as the abolishment of the "issuance eligibility criteria" for bond issues in Japan by non-residents, and resident issues of Euroyen bonds outside Japan.

To summarize, due to the liberalization and deregulation of access to the Euromarkets as well as to Tôkyô offshore markets during the last twenty years or so, large Japanese firms in particular experienced a dramatic increase in financing alternatives.

2.2.2 Financial Liberalization: The Domestic Level

Internationalization, as stated above, had an immediate effect at the domestic level. Particularly in the Euromarkets, issuing requirements of bonds were already much more relaxed, which substantially increased the number of eligible firms and

---

3 For a detailed chronology of the internationalization process of the Yen, see Hall (1998), pp. 71-74.

4 The pressure from international markets on Japan's domestic market which quickly gained momentum after the revision of the FEFTCL, has been well underlined by Ramseyer (1994). He provides evidence showing that between 1980 and 1984 foreign bond issuance increased from 683 billion to 2.8 trillion yen while the domestic bond market increased from 1.1 trillion to 2.3 trillion yen.

strongly reduced issuing fees (see also excursus on bonds in 2.3.3). Together with a broad range of financial instruments also available in the Tôkyô offshore market, which were not (yet) available in domestic Japan, there was no choice but to gradually proceed with deregulation and liberalization of domestic financial markets as well. Any other decision would have led to an immediate (further) decline in Japan's attractiveness as a financial center of worldwide importance.

Furthermore, interest rates on deposits were liberalized throughout the 1980s, and this subsequently raised banks' financing costs, which resulted in a decline in banks' profits, i.e., interest rate margins (Bank of Japan 1996). The easing of regulations on deposit rates started with the introduction of certificates of deposit (CD) in May 1979. During the 1980s, the minimum amount needed to hold a CD was continually lowered. Full-scale liberalization of interest rates was started in 1984, when the Japanese government placed the deregulation of interest rates on the agenda for the U.S.-Japan Yen-Dollar Committee[5]. With liberalization accelerating, the administration increasingly allowed additional, innovative types of deposits (e.g. money market certificates (MMCs) in May 1985), of which interest rates moved with market rates and which reached a high degree of acceptance as alternative products for deposits. By 1993, with the exception of small-denominated and demand deposits, all bank deposit rates had been liberalized. In October 1994, all remaining regulations on deposit interest rates were finally abolished, except for those on "current deposits" (*tôza yokin*)[6].

Concerning the liberalization of the domestic banking sector, among numerous amendments, we can identify two reforms that had an exceptionally strong impact on the composition of the financial system. First, the Banking Law introduced in April 1982 enabled banks to trade with government bonds. Second, and even more important, the Financial Reform Act introduced in April 1993 eliminated the divisions between commercial banks, trust banks and securities houses[7]. Until 1993, Japan's banking structure was segmented and categorized along various lines: (a) between commercial banks and securities houses (along this line, trust banks developed as a separate category; their main function was to manage funds entrusted to them on both corporate and private accounts); (b) between functions (by size, location and/or profession of the customer); and (c) between short-term versus long-term lending among large banks (Zenginkyô 1994; Schaede 1996). However, since 1993 banks and securities houses have been allowed to move into each other's primary market through wholly owned subsidiaries, thus triggering competition among financial intermediaries. All in all, we

---

5 See Frankel (1984) for detailed explanation of the agreement of the Yen-Dollar Committee.

6 For a detailed chronology of the liberalization process of the Japanese money market, see Hall (1998), pp. 90-91, 94-97.

7 For a detailed chronology of cross-entry by financial institutions under the Financial System Reform Act of 1993, see Hall (1998), pp. 128-129.

can state that rates on bank loans as well as deposits have become increasingly market-driven. Note, however, that Article 65 of the Securities and Exchange Law still prevents any single firm from engaging in both commercial and investment banking, meaning that the new regime is far less liberal than what is found in major European countries, where "universal" banks can offer all financial services under one roof.

The ability to engage in stock futures and options trading, first undertaken on the Osaka Securities Exchange in June 1987 (although stock index trading was not introduced until 1988), further increased the variety of products available in domestic financial markets (e.g. Schaede 1990; Hall 1998).

2.2.3 Excursus: Bank Versus Bond Financing

With regard to corporate financing and subsequently bank-firm relationships, the most important aspect of the liberalization process probably has been the deregulation and innovation of Japan's corporate bond market[8]. As we will see in chapter 2.4, it is the foundation that finally allows firms to choose between the monitored debt typically supplied by commercial banks, and arm's-length debt obtained in corporate bond markets, in order to find a financial mix that best suits each individual firm.

Until the liberalization of bond markets for Japanese firms, "specialized financial institutions, operating in well-defined product markets ... limited financing alternatives for the various corporations. In addition, a lack of price differentiation among products reduces the incentives for participants to break these relationships" (Meerschwam, 1991: 135). Moreover, from the mid-1930s until 1979, the ability of Japanese firms to issue corporate bonds for the domestic market had been rigorously restricted[9] by the principle of collateral and eligibility requirements (Okuda and James 1990; Hodder and Tschoegel 1985, 1993; Horiuchi 1996). Analogously, until the early 1980s legal restrictions limited bond issuing by Japanese firms in international markets, thus making it hard for Japanese firms to circumvent the regulated domestic bond market (Hoshi 1995). As a consequence, regulations prevented the overwhelming majority of firms – even many of the safest corporations – from issuing bonds. It hindered the development of a flexible price mechanism and thus further strengthened Japan's

---

8 For a detailed discussion of different types of corporate bonds issued by Japanese firms, see e.g. Beason and James 1999, pp. 61-64.

9 The partial approval of overseas bond issuance in 1973 (the funds raised were to be used abroad), and the following changes which fully approved overseas bond issuance in 1974 (the funds could also be used domestically) effected only a very limited number of firms like NTT and electricity firms, which were most likely public. It did not have any significant effect on the overall structure of corporate financing of the private sector.

bank-dominated financial system[10]

Observing this period, it is interesting to note that in contrast to standard theory, the underdeveloped nature of the corporate bond market did not seem to be an obstacle to rapid industrial development during Japan's high-growth period. Obviously, the intimate relationship between banks and firms worked sufficiently well to help industrial sectors finance their large investment expenditures (Horiuchi 1996). Yet, the "historically heavy use of bank financing by Japanese firms is in part a regulatory phenomenon because alternatives to bank financing did not exist for the vast majority of firms prior to the 1980s" (Anderson and Makhija 1999: 313). In other words, Japanese firms did not necessarily find bank loans more advantageous; they rather had fewer options.

However, in the early 1980s the internationalization of financial markets exerted great pressure on domestic bond markets[11]. After the revision of the FEFTCL in late 1980, particularly the Euromarkets became very attractive for large Japanese firms. In comparison with Japan's domestic bond markets, the Euromarkets were much more competitive in terms of criteria for determining the eligibility of issuers, collateral requirements, terms of issue, and issuing procedures. Hence, Japanese authorities had to act against an increasing hollowing-out of domestic markets. This led to a gradual relaxation of restrictive domestic bond issuing criteria during the 1980s. Furthermore, the revision of legal restrictions contributed to an overall deepening of the market by broadening the range of bond types on offer, diversifying their maturities and diversifying selling methods[12]. Since the beginning of 1993, the Corporate Bonds Act has been further eased several times by simplifications to the bond issuing process, substantially reducing the cost of bond financing and thus even further increasing the number of firms eligible to issue both secured and unsecured domestic bonds. Finally, in January 1996 the issuance eligibility criteria were abolished entirely.

Each step of further liberalizing the domestic bond market has sparked a flurry of bond issues, and prospects for the development of a deep, liquid secondary bond market are better than ever (Choy 1994). Moreover, banks, which

---

[10] Underlining the monopolistic position of banks in this system, Ramseyer (1994: 237) even speaks of a Japanese "bank-loan cartel", which successfully locked out any competition from the bond market.

[11] In 1979, the value of corporate bonds issued by Japanese firms in domestic markets totaled over Yen 2.4 trillion, more than three times the value of offshore issues. In 1985, however, total Japanese corporate bond issues offshore had risen by Yen 3.3 trillion, more than 25 per cent greater the value of domestic issuing (Association of Securities Underwriters: *Monthly Report of Japanese Bonds*, various issues).

[12] It is interesting to note, however, that the liberalization of the domestic market does not seem to have mitigated the hollowing-out process effectively. In fact, since 1990 the importance of offshore markets for Japanese corporate bond issuing is further increasing.

previously had a strong influence on a bond issue's terms, are now often limited to custodial duties. This allowed a wider variety of non-bank financial firms to begin advising clients on bond issues. With increasing competition, banks, like brokerage houses, were forced to begin negotiating underwriting fees to retain customers[13]. As a consequence, the bond-underwriting cartel (*kisai uchiawasekai*, later named *kisaikai*), which had controlled the domestic bond market since 1933, broke down[14].

However, compared to American and European bond markets, where the variety of available products enables firms to tailor bond issues closely to their individual needs, the variety of products in the Japanese bond market is still limited. Furthermore, Japanese investors still tend to hold bonds to maturity rather than trade them freely like stocks. Hence, comparing the variety as well as liquidity of Japan's bond market to that of its European and American counterparts, the Japanese market still seems rather immature.

Though the Japanese bond market still has much room for improvement, we can say that overall, with the increasing liberalization of the corporate bond market, Japanese firms have a much wider variety of capital sources at hand. Hodder and Tschoegl (1993: 158) note that "at least some firms are finding securities issues cheaper and/or less burdensome than borrowing via loans". They substantially increase the amount of bond issues and correspondingly decrease their reliance on bank loans (Hoshi 1995). Thus, one can state that especially during the second half of the 1980s, this process led to a massive "securitization" of corporate finance (Horiuchi 1996).

Not surprisingly, along with the liberalization of legal restrictions, the soaring Japanese stock market during the latter half of the 1980s further helped to increase the popularity of bond issuing[15], especially that of convertible bonds[16] and bonds with warrants[17]. The overall expansion in bond financing was furthermore accelerated by the fact that in the second half of the 1980s, the nominal issuing cost of convertible bonds and bonds with warrants declined sharply. In addition, the Bank of Japan (1996) survey points out that during this

13 According to Semkow (1992), during the period when the underwriting cartel controlled Japan's corporate bond market, collateral management fees as well as commissions on domestic bond issuances were, on average, 15 times higher than commissions on similar issuances in the Euromarket.

14 See Kôshasai hikiuke kyôkai (1980) on the history of Japan's bond market.

15 A further boost to the with-warrant bond market was provided during fiscal 1995 with the sanctioning of public offerings for firms whose shares were registered with the Japan Securities Dealers' Association (Hall 1998).

16 Convertible bonds are bonds that can be converted into stocks of the issuing firm at a stated time in the future.

17 With-warrant bonds provide the holder with the right to purchase a given amount of stock at a fixed price at a stated date(s) in the future.

period, corporate managers seemed to have perceived the issuing cost to be only the coupon rates of these bonds and thus judged the level of cost to be extremely low.

Note, however, that an important characteristic of the liberalization of Japan's bond markets is the advantageous treatment of large, well-established firms like those in our sample (see chapter 3), rather than of small- and medium-sized firms. To give an example, during the period of the surging stock market in the second half of the 1980s, convertible bonds and bonds with warrants were popular among large, listed firms. However, firms not listed at the stock exchange could not take full advantage of the liberalization and deregulation process at that point. Regarding this fact, Horiuchi (1996) speaks of a distorted liberalization. Even if the still-incomplete reform of the institutional and legal framework that locks out many small- and medium-sized firms up to the present date develops further to also include smaller firms, large, listed firms will continue to enjoy additional advantages of the corporate bond markets. In contrast to small, unknown and/or unlisted firms, large firms have most likely built up a good reputation over many years and listing requirements of the stock exchange as well as the Ministry of Finance secure the highest and most standardized degree of disclosure in terms of Japanese standards. Thus listed firms tend to have a lower agency cost potential than small emerging firms. Considering the fact that firms with a low agency cost potential (suggested by a higher degree of disclosure together with larger size, higher profitability, lower operating volatility, lower growth opportunity, and larger amount of collateralizable assets) are more likely to depend on corporate bond debt as opposed to monitored debt supplied by specialized commercial lenders (see Berlin and Loeys 1988, Diamond 1991, Rajan 1992, and Chemmanur and Fulghieri 1994), the advantages for large reputable firms become even more obvious.

Finally, it is interesting to note the changes in the distribution of corporate bondholders. Up to the mid-1970s, Japanese banks were by far the most important holders of corporate bonds. However, bond holding by banks substantially declined during the following two decades, while the household sector has substantially improved its position, now being as important as the banking sector (Bank of Japan 1996). In a breakdown of investors for the new issues in 1985, Nihon Keizai Shinbun Sha (1987) reports that the largest share was purchased by individuals (49.1 per cent), while financial institutions bought only 28.5 per cent. This development underscores that the decreasing power of banks to control the corporate bond market is not only evident from the supply side but also from the demand side.

2.2.4 Most Recent Developments and Outlook

Newer developments show that the process of liberalization and deregulation have continued during most recent years and will reach into the next century. In November 1996, then-prime minister Hashimoto announced that under his government, a new and even more powerful era of financial deregulation would be introduced and would result in completely "free, fair, and open" financial markets by March 2001 (see Yatsunami 1999; Baron 2000; among others). This effort, which all successive cabinets promised to continue, was (roughly) modeled on the 1975 Market Day reforms in the United States and the 1986 Big Bang in the UK. It is supposed to lead to the final stage of deregulation in order to bring Tôkyô's financial markets up to par with those of New York and London.

Among the most prominent reforms implemented since Hashimoto's announcement[18] are: the 1997 Bank of Japan Law, through which the Central Bank gained a greater measure of independence from the Finance Ministry; the revamped FEFTCL as of April 1, 1998, which further liberalized cross-border capital transactions; and a widening of the scope of financial instruments in fiscal years 1997 and 1998, introduced by several amendments to the Securities Investment Trust Law and Securities and Exchange Law, as well as by new interpretations of the Commercial Code.

Up to the present date, the media-effective announcement of a Japanese "Big Bang" has not substantively fulfilled hopes for strong measures that will trigger an abrupt shock wave of financial liberalization. Strong pressure from various power groups such as banks, as well as the recent recession, seem to prevent most drastic changes. Some issues, notably the postal savings system, remain completely unresolved. Overall, however, Japan's financial system has undergone major reforms during the last twenty years, and as we will see in more detail below, the deregulation of Japan's financial markets already has had wide-ranging impacts. The financial system is far more market-oriented than it was during the high-growth period from the early 1950s to the beginning of the 1970s, as well as the decade that followed thereafter when banks largely dominated the financial system. Even though some domestic pressures still act against further liberalization, international pressures seem to be strong enough to further push Japan towards a more market-oriented and internationally competitive financial market structure in the long run. Since the internationalization of Japan's financial markets, it is more a discussion of how, rather than whether, liberalization and deregulation will further advance. In fact, the more competitive banks already recognized these developments several years ago and thus successfully have tried to redefine their position in the Japanese financial system.

---

[18] For a more detailed account on the recent reforms: Ministry of Finance, Monthly Finance Review, No. 289, August 1997, and No. 298, May 1998.

2.3 The Boom-Bust Cycle

Next to the reforms of the legal and institutional framework of Japan's financial markets, the causes and consequences of the latest economic boom-bust cycle had an additional effect on the financial structure of firms, thus also contributing to the changing position of banks in general and the role of bank-firm relationships in particular. Since there has already been substantial work on the boom-bust cycle (e.g. Hall 1998), we will only give a brief overview of the cycle's development and limit our discussion to the most prominent features that are particularly important to understanding the framework in which the changes of corporate finance and subsequently the formation of the banks' new position have been taking place.

2.3.1 Building up the Bubble

Within the process of financial deregulation, one of the most prominent characteristics of Japan's financial markets in the latter half of the 1980s was a rapid increase in fundraising with low interest rates through capital markets (see also chapter 2.4). Figure 2-1 shows the financial position of Japan's corporate business sector since 1970. We can observe that the exceptional development during the second half of the 1980s caused a "debt bubble" (Shimizu 1997a: 25). The dramatic increase of asset prices, especially real estate and stocks, finally led to an "asset bubble" (Ito and Iwaisako 1996: 144). Main economic indicators responded accordingly: from the second quarter of 1987 until the third quarter of 1990, there was a boost in domestic demand, and output responded very strongly. The real GDP growth rate was close to or even exceeded five per cent from 1987 to 1990, and the unemployment rate remained very low.

As will be discussed in more detail below (see chapter 2.4), the increase of funds was mainly based on long-term fundraising from the capital markets, which accompanied a reduced share of borrowings from financial institutions (mainly banks). This development can be observed especially among large firms, which have easy access to the capital markets. The share of bank-borrowings (total of short- and long-term loans payable) contributing to total liabilities declined from 31.0 per cent in fiscal year 1980 to 19.9 per cent in 1994, and has remained roughly at this level ever since (MOF: Financial Statements of Incorporated Businesses, yearly issue).

Figure 2-1: Financial position of the corporate business sector, 1970-1998 (times nominal GDP)

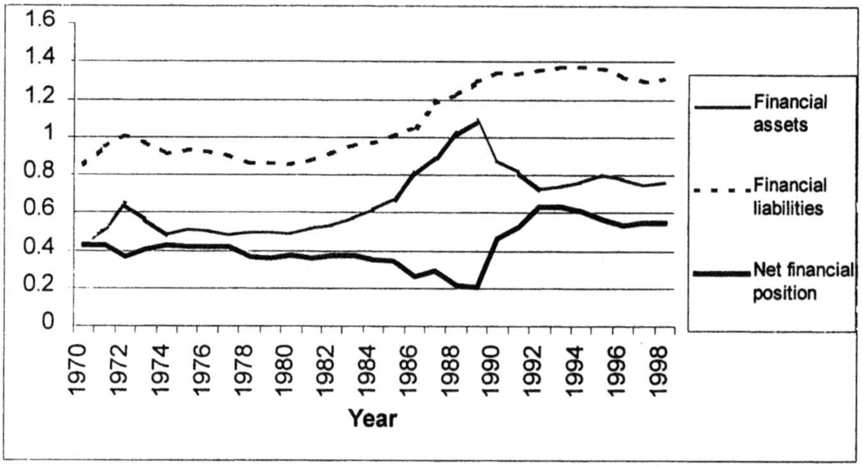

Notes: 1. Source: Flow of Funds Accounts, Bank of Japan

2. Net financial position = Financial liabilities - Financial assets

3. Financial assets exclude 'trade credits', 'foreign claims' (foreign trade credits, foreign direct investments).

4. Financial liabilities exclude 'stocks', 'trade credits', 'foreign debts' (foreign trade credits, foreign direct investments and other foreign debt <excludes foreign bonds>) and 'others'.

The increased fundraising was favored by new opportunities for the liberalized financial markets to access new capital (as mentioned above), and to issue equity when stock prices were very high. Corroborated by exceptionally low interest rates and soaring stock prices at that time, many firms raised more funds than they needed for their capital investment. While increasing their debts, they invested in assets, among these stocks and real estate, which had very high prospective rates of return.

The increase in the firms' stock investment using equity-financed funds further facilitated the firms' equity financing and then led to spiraling stock prices as follows: Stock investment by firms using equity financed funds, particularly evident among large manufacturing firms, increased and pushed up stock prices mainly of non-manufacturing firms. This in turn raised average stock prices and strengthened the expectations of overall stock increase, thereby further facilitating equity financing by firms and further accelerating the rise in stock prices (Bank of Japan 1996). Nevertheless, theoretically speaking, such a spiral involves some contradictions; market participants, however, were not concerned about these contradictions, and based on the entire postwar experience "everybody expected that land prices in Japan would never fall" for any sustained period (Shimizu 1997a: 34). Large amounts of funds were raised through equity financing with little concern for negative effects on future stock prices. Deregulation favored this process by reducing the cost of raising funds when increased channels of investment offered returns higher than the interest costs of the funds raised.

During this period, the position of banks changed dramatically. In fact, the bank's reaction to the changing environment further accelerated the bubble. Many large firms started to raise large amounts of long-term funding from the capital markets, which reduced the amount and profitability of bank loans to large firms. In order to keep large firms as their customers, banks were forced to accept transactions with very low profitability. Looking for new and more lucrative alternative markets, while being confronted with an increasing number of competitors due to deregulation, banks increased their loans to medium- and small-sized firms and the real estate industry. These sectors still had limited access to the capital markets; hence banks could secure relatively large lending margins from these customers. However, by shifting to these new customers, banks had to accept an increase in risk. In retrospect, the expansion of such lending failed to generate profit margins sufficient to cover the substantial risk involved: As will be further discussed below, when the risk materialized as asset prices fell, banks were left with large amounts of non-performing assets (Bank of Japan 1996). Furthermore, banks believed that they had to increase the scale of loans in order to reduce operating cost ratios. The realization of this strategy was made possible by securing loans with land as collateral, which had an ever-increasing value due to soaring real estate prices. As a result, banks contributed heavily to the debt bubble.

2.3.2 The Bubble Burst

After stock and land prices (but also assets like real estate, art, and country <golf> club memberships, etc.) doubled and tripled in a matter of a few years during the second half of the 1980s, many asset prices plummeted in the first half of the 1990s. To give some examples, the Nikkei 225 index was 11,542 in December 1984, climbed up to 38,915 in December 1989, and plummeted to 15,951 in June 1992. The Commercial Land Price Index for large cities increased from 38.4 in 1986 to 103.0 in 1991, and decreased to 71.4 in March 1993.

This process was largely influenced by a self-correcting mechanism inherent in stochastic speculative processes, and a change in the conduct of monetary policy in this period. After monetary authorities became fully aware of asset price inflation, the Ministry of Finance introduced several measures to slow the rise of land prices. More importantly, in May 1989, the Bank of Japan decided to conduct a tight monetary policy. As a result, the short-term interest rate declined. However, somewhat surprisingly, the long-term rate remained stable in the beginning. It seemed that the market thought that the Bank of Japan policy change was only temporary. This belief may also be a reason why stock prices continued to rise until the last trading day in December, 1989. Thereafter, however, the long-term rate quickly caught up with the short-term rate, after the market became convinced that monetary policy ought to be tightened. Within a year, the term structure was back to the status quo: during a period of monetary contraction, the short-term rate was again higher than the long-term rate and when the official discount rate was decreasing, it was lower.

Due to the severe decrease in asset value, the bank strategy of increasing scale while securing the loans with collateral of land collapsed, and with it, the belief in the stability of the Japanese financial system. Trading in the real estate market became very inactive and land did not work as collateral to repay loans or even cover losses. It became apparent that using land and real estate as collateral was a good means to secure individual bank loans when taken separately. However, its function as collateral is very restricted in the case of large bad loans caused by severe depression due to the fall in market values and limited liquidity (Shimizu 1997b). As a result of this high-risk strategy, large amounts of unrepayable bad loans accumulated and led to the collapse of several banks and other financial institutions (see e.g. Schaede 1996), and exerted pressure on the rest of the financial system. The deterioration in the quality of loans to the real estate sector (the so-called *jûsen*[19] problem) was the primary problem. It was compounded by a drop in the value of the banks' large equity holdings and a prolonged recession, which further led to growing loan problems. The non-performing loan problem together with the Bank of International Settlement adequacy ratio[20] caused banks to be very sensitive to loan risk. Banks reduced loans, and the money supply dropped to very low levels, even showing negative values in 1992 when the Bank of Japan did not use adequate power to control the money supply. Note, however, that the banks' reduction of loans largely affected small- and medium-sized firms rather than large firms like those we will focus on in the following analyses (see chapter 3).

2.3.3 The Recession Phase

The decline of asset prices and subsequently the virtual collapse of the financial system triggered Japan's most severe postwar recession, which shows

---

19 *Jûsen* or housing loan companies are non-bank subsidiaries of financial institutions specializing in housing loans.

20 In 1988, when asset prices were high, Japanese banks were allowed to count 45 per cent of their equity holdings as part of their tier II capital to meet the Basel risk-based capital ratio. The value of these equities were devoted to meet capital asset requirements, resulting in an increased cost of capital, reducing the incentives of banks to make loans, and contributing to the drop in credit expansion.

only a few tentative signs of recovery in the late 1990s (Itami 1998, Albach 1999a). Consumption fell as households saw the real value of their equity and real estate holdings decline. Consequently, GDP went sideways and finally declined, being reinforced by depressed demand in export markets. Unemployment rose slowly to the highest level in Japan's post-war history. This prompted a fall in fixed investment, which was further weakened by an overhang of excess capacity accumulated during the asset inflation phase and the credit crunch induced by the decline in "hidden" bank capital (Cargill, Hutchinson and Ito 1997). The depression was further aggravated by the rapid contraction of the money supply and the slow reaction of policy measures to increase money supply and re-activate real estate transactions.

Reacting to the non-performing loan problem, during the first phase of the recession, regulatory authorities introduced several measures that all had to be abandoned later on, since they were based on the belief of a quick recovery of the economy in general and the real estate market in particular. During the second phase of the recession, when the financial crisis unfolded in full, Japanese authorities introduced new measures to solve the problem. Most prominently, for the first time in the post-war period, the Ministry of Finance closed several financial institutions, among these banks and securities companies as well as all seven *jûsen* companies. To bridge the closings, the Tôkyô Kyôdô Bank was established in March 1995 to receive the remaining assets of failed smaller institutions. The Bank was reorganized in September 1996 into the Resolution and Collection Bank, which was loosely modeled after the U.S. Resolution and Trust Corporation. However, it is important to note that, although a considerable number of banks as well as other financial institutions fell during the 1990s, Japanese authorities made sure that they all landed softly. Thus, no depositor lost any money[21]. Even though these measures were taken to keep the level of lost confidence in the banking sector low, banks still suffered a strong loss of reputation in Japan as well as abroad[22].

In order to further consolidate the financial system, the MOF and BOJ encouraged mergers of financial institutions, of which the latest and biggest will be that of Fuji Bank, Dai Ichi Kangyô Bank (DKB), and the Industrial Bank of Japan (IBJ) in the year 2000. Furthermore, the MOF allowed massive write-offs by several banks. Further mitigation of the non-performing loan and *jûsen* problem was attempted by a massive infusion of public funds into the private financial sector, though with limited success, due to the disclosure of ever more non-performing loans.

Another attempt to prevent the financial system from a collapse is a variety

---

21 In order to ensure credibility of the financial system, in summer 1995 the Ministry of Finance announced that *all* deposits would be guaranteed until March 31, 2001.

22 The so called "Japan Premium" which had to be paid by Japanese banks in international financial markets during much of the 1990s is just one sign for the lost reputation and the uncertainty that was felt about the soundness of Japanese banks.

of "gimmicks" (Lincoln 1998) that do not, by any means, seem to solve the crisis in the long run, but that make the administration seem to help in the short run. Among the most striking are attempts to improve balance sheets to prevent public recognition of weakness or insolvency. Introduced in 1998, financial institutions have to revalue real estate holdings, mainly land that they have already held for decades, to market value, while it is optional to do the same with stock portfolios, of which many shares were purchased near the peak of the stock market. The rising mistrust in the soundness of Japan's financial system at home and abroad has been further reinforced by various scandals that have emerged in the 1990s, revealing shocking examples of indiscretion and malfeasance (see e.g. Albach and Miarka 1997).

Domestic demand has deteriorated further and due to the last Asian crisis, external demand was also declining, while at least another $ 50 billion was added to the $ 630 billion of domestic non-performing loans acknowledged by the MOF in late 1997. Apparently, the latest recession is the most severe not only in magnitude but also longevity. The magnitude of the asset price inflation and the subsequent deflation has drawn particular attention to the relationship between macroeconomics and financial markets.

2.3.4 Excursus: Mechanisms of the Cycle and the Role of Assets

To explain the main features of the mechanism behind Japan's recent boom-bust cycle, we can largely follow the basic structure of the dynamic model by Kiyotaki and Moore (1997). Most aspects of this model are remarkably well suited to explain financial factors and economic fluctuations in the Japanese context, where collateral plays an outstanding role. Based on the incomplete contract approach, in this model the endogenous fluctuations in the market price of an asset, which is used both as input in production and as collateral for lending, are the key factor for changes in the borrower's net worth.

The mechanism functions as follows: Starting from the steady state, a negative shock on the real estate equilibrium prices is assumed. This immediately affects the price given by a no-arbitrage condition, at which the asset will be resold in the next period. First, since the value of assets predisposed to be used as collateral decreases, financial constraints become more severe. Thus less profit will be made in the productive activity. Second, more profit will be made simultaneously on the rental activity. Nevertheless, it turns out that the first effect dominates. Furthermore, the negative shock on the real estate equilibrium price causes an increase in the real estate price at the next date.

Summing up the basic features of the model, the cyclical movement can be explained as follows: if the asset price is larger than it is in equilibrium, the second effect dominates and the price at which the asset will be resold becomes smaller than in equilibrium. Reaching a price smaller than in equilibrium, which causes the first effect to dominate, the process starts over in the same pattern already described above. It is important to underline that for this result of pro-cyclical deterministic dynamics of land prices, it is crucial that the demand for a productive asset is an increasing function of its price, given the case that firms are credit constrained. Hence, asset price fluctuations incite shocks in spending and production, and net worth distribution propagates over time and sectors. This leads to persistent fluctuations in output and asset prices. Note that the mechanism corresponds to Fisher's (1933) debt deflation theory, that changes in asset values are the principal means by which financial forces lead to economic decline.

Considering the empirical evidence available, the initial impulse that triggered the asset inflation phase in 1985/86 was most likely related to changes in fundamentals. At this time, Japan recorded strong aggregate productivity gains, which can be interpreted as a productivity shock. With it, commencing in the Tôkyô area, the demand for real estate increased significantly. In the following years, the demand for land increased even further, giving rise to an additional appreciation of its value. Moreover, monetary policy decisions pushed interest rates to very low levels, thus further supporting the demand for real estate.

Due to rising land prices and decreasing interest rates, firms were able to borrow more as the value of their collateral increased substantially, thus triggering a debt and investment expansion. Moreover, stock prices reacted immediately to the positive shock and they appeared to be leading land prices, even though they incorporate the rising value of land held by the firm, thus increasing with land prices. This can be explained by the fact that the stock market can react and anticipate more quickly than the real estate market.

Following the model of Kiyotaki and Moore (1997) further, prices finally rose much beyond the equilibrium price of assets until the price effect, corroborated by the sudden rise of interest rates, triggered a negative shock on the real estate demand. As a result, the price of assets intended to be used as future collateral decreased, and so did the price of stocks. Here again, the reaction of stock prices came somewhat more quickly than that of real estate prices, due to larger imperfections of the real estate market. The decreasing prices of assets intended for use as collateral caused financial constraints, which further increased while asset prices remained low or decreased even further. Consequently, debt and investment of firms followed the negative trend and GDP went sideways and finally declined.

In their model, Kiyotaki and Moore (1997) predict that the asset prices will go up again as soon as asset prices are significantly under the equilibrium price of assets. A positive shock would further help to trigger the next phase of the cycle. Apparently, the financial policy currently conducted in Japan prevents asset prices from quickly reaching this level, therefore prolonging the recession of the Japanese economy, which already has had a negative effect on the world economy.

2.3.5 Conclusion

Legal and institutional changes since the 1980s have already substantially weakened the monopoly power of banks and thus had a strong impact on the role of bank-firm relationships (chapter 2.2). Considering the development of the economic framework in Japan during the boom-bust cycle, we can identify at least two additional developments that further weakened the position of banks in a relationship with large firms. First, the rising stock market during the boom period helped especially large firms to use sources of finance other than bank debt, thus exposing banks to strong competition. As we will see in chapter 2.4, once firms had started to use arm's-length finance such as bonds as a major source of capital, many firms continued to do so even after the end of the boom period.

Second, during the bust period, the banking sector had to take the strongest hit. After land prices started to plunge dramatically and land could not serve sufficiently as collateral any more, a large number of banks were substantially weakened and some even had to be shut down. Even after almost ten years, the majority of banks have not been able to overcome this loss of financial strength as well as reputation. The struggle to survive had consumed many resources that could have been used otherwise to increase competitiveness. Furthermore, since the early 1990s demand for finance strongly decreased, while investment opportunities declined, thus further increasing competition among the various available financial sources. Summing up, one can state that concerning bank-firm relationships, the developments during most of the boom as well as the bust period encouraged especially large client firms to substantially loosen relationships with their banks.

2.4 Changes in Corporate Finance

Most interesting for our analysis is that, as a result of liberalization and deregulation as well as overall economic development during the boom-bust cycle, many Japanese firms started to choose more freely among alternative sources of debt. To derive the largest amount of information possible about the firms' debt structure, in the following the debt composition of large Japanese firms of the manufacturing sector over the last two decades will be examined, since it is the most useful piece of financial mix information available[23]. Size and

---

[23] For an analysis of changes in the corporate balance-sheet structure of large non-manufacturing firms as well as small and medium-sized firms between 1980 and 1994, see Bank of Japan (1996).

mandatory listing of the sample firms[24] that contribute to the aggregate data used in the following ensures that these firms have access to the full range of possible sources of debt. Having full access, the change in the financial mix structure is likely to reflect relative supply conditions. One would expect that the relative contribution of informed capital (bank financing) and arm's length finance (especially corporate bonds) should change in a systematic way when the supply of these two sources of debt are influenced by a positive or negative shock. In fact, the two most significant changes in the financing of large manufacturing firms during the last two decades were (a) a higher reliance on internal funds; and (b) a decline in the share of borrowings from banks in external funds. Table 2-1 and figure 2-2 show the process of disintermediation evident in the balance sheets of our sample firms more closely: Table 2-1 reports ratios of liability and equity subtotals to assets (mean values) for the fiscal years 1979, 1982, 1985, 1988, 1991, 1994, and 1997. While book leverage declines from 70.06 per cent of assets in 1979 to 55.54 per cent of assets in 1997, bond financing almost quadrupled from 3.24 per cent of assets in 1979 to 12.88 per cent in 1991, and has maintained a level of around 11 per cent ever since. As expected, bond issuing rose during the time of a booming stock market. The stock market, which appeared to be ever rising during the second half of the 1980s, made convertible bonds and especially bonds with warrants very popular. The development of bonds with warrants underscores this view. Rising until the early 1990s and then strongly declining over the rest of the 1990s, the popularity of this type of bonds seem to have closely followed the boom-bust cycle of the stock market[25].

However, even more interestingly, bonds in general and convertible bonds in particular managed to keep their attractiveness despite the burst of the Japanese stock market bubble. In other words, for large manufacturing firms, the high ratio of bond financing to total assets is not a short-lived phenomenon that went along with a soaring stock market. After liberalization, bonds rose to be a source of capital that seems as important as bank finance. And despite the weak performance of the stock market, this has continued throughout the 1990s.

The increase of bond financing offset a decline in bank loans. The ratio of loans (short- and long-term) to assets, which reached 22.32 per cent in 1979,

---

24 See chapter 3 for an in-depth description of the sample firms.

25 Note that bonds with warrants (originally undetachable) were legalized not until 1981, and the detached warrant components were not tradable in Japan until 1986. In addition, until March 1994, the so-called "combined accounting method" was applied to bonds with warrants, just like to straight bonds. In this accounting method, subscription prices of warrant bonds were not separated into the prices of bonds and warrants in accounting terms, regardless of whether warrants were detachable or un-detachable. Hence until this method had been replaced by the "separation method" in April 1994, the seemingly low cost of bonds with warrants may have further contributed to the rising popularity of bonds with warrants in the past (see Bank of Japan 1996 for a detailed description of accounting practices).

declined to 10.67 per cent in 1991 and recovered only slightly during the recession period thereafter. In other words, until the 1980s bond financing could only be used by a very exclusive number of firms, thus being virtually of no significance for the majority of firms. However, after the legal framework of the bond market was revised during the early 1980s, the ratios of loans to assets and bonds to assets converged, finally taking about the same weight in the composition of firm debt during the late 1980s and throughout the 1990s.

Next to the changes concerning the sources of firm debt, we furthermore have to indicate strong changes in the term-structure of firm debt. Table 2-1 shows that during the last twenty years, the percentage of current liabilities to total assets has declined by close to fifteen per cent, while the share of long-term liabilities remained rather stable.

Table 2-1: Ratio of liability and equity subtotals to assets (mean values) for large Japanese manufacturing firms in per cent, 1979-1997

Balance Sheet	Fiscal year						
items	**1979**	**1982**	**1985**	**1988**	**1991**	**1994**	**1997**
Trade credit	20.35	21.33	21.34	16.63	17.08	13.31	15.38
Short-term loans	13.81	10.58	11.99	11.31	7.54	8.88	8.80
Other current liabilities	17.52	17.38	14.71	13.14	13.78	12.43	12.82
Total current liabilities	51.68	49.29	48.04	41.08	38.40	34.62	37.00
Long-term loans	8.51	7.70	5.69	3.67	3.13	5.62	3.95
Bonds	3.24	3.02	6.32	10.38	12.88	11.81	10.79
Straight bonds	1.83	1.39	1.52	2.65	3.45	3.63	3.93
Convertible bonds	1.41	1.63	4.50	5.80	5.81	6.26	6.80
Warrant bonds[25]	0.00	0.00	0.30	1.93	3.62	1.92	0.06
Other long-term liabilities	6.63	5.66	4.04	4.15	3.91	3.93	3.80
Total long-term liabilities	18.38	16.38	16.05	18.20	19.92	21.36	18.54
Total liabilities	70.06	65.67	64.09	59.28	58.32	55.98	55.54
Stock	5.71	5.93	7.60	9.32	10.18	11.03	10.94
Retained earnings	24.23	28.40	28.31	31.40	31.50	33.00	33.52
Total equity	29.94	34.33	35.91	40.72	41.68	44.02	44.46

Source: KAISHA database, sub-sample of 92 firms[26]

[26] For a detailed description of the database and the sub-sample used throughout the analysis, see chapter 3.

Figure 2-2: Development of the firms' debt balance; growth rates between 1980-1998

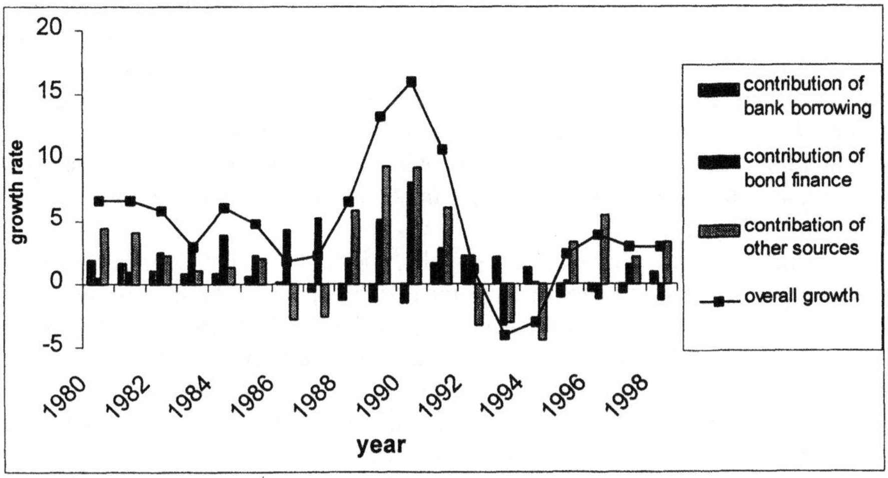

Source: KAISHA database, sub-sample of 92 firms

Analyzing the contribution of different sources of debt to the overall growth of firm debt may even more explicitly reveal the disintermediation process that has been taken place during the last two decades. Figure 2-2 shows to what degree financing by bank loans, bonds, as well as other sources of debt (like trade credit etc.) contributes to the development of the firms' debt balance during the years 1980-1998. As soon as access to bonds started being liberalized in the early 1980s, bond financing began to contribute increasingly to the growth rate of firm debt. And even when the overall growth rate experienced a short-lived decline between 1985 and 1987, the contribution of bond financing continued to increase. Initially caused by increasing investment opportunities during the late 1980s, the demand of funds then sharply increased. This period goes along with the liberalization phase already outlined above, in which capital controls and lending rates where deregulated and a money market was established, thus triggering a positive shock in debt supply. Note that rising asset prices during the period between 1987-1990 led to a sharp increase in the value of firm assets used as collateral, and subsequently to an increase in firms' net worth. This allowed many firms to raise large amounts of informed and especially of uninformed capital (see also chapter 4).

The development of the different sources of debt shows that the overall increase of debt supply was especially supported by an excessive use of arm's-length finance, which practically exploded after long being suppressed. In 1986 and 1987, the demand for bonds was extremely strong, even slightly substituting the negative contribution of all other sources to the growth of firm debt. This development was mainly due to the introduction of additional deregulation in the second half of 1985 and in 1986, which further eased access to the bond market. During the following years, until 1991, bond financing continued to contribute strongly to the overall growth of debt. As has been noted above, next to the deregulation of bond markets, the attractiveness of bonds was furthermore supported by a sharp rise of collateral value, which contributed to the increasing firm value. Not surprisingly, the growth rate of bonds, in large part due to the strong demand for convertible bonds and bonds with warrants, reached its highest point in 1990, after the securities market had climbed to its all-time peak on the last day of the preceding year.

While bond financing was very popular, bank debt did not contribute at all to the strong growth rate of debt of large manufacturing firms throughout the boom period. Obviously, monitored finance was perceived to be more costly compared to arm's-length debt. When the Basel risk-based capital adequacy ratio was negotiated in 1988, it implied an increase in the cost of capital, thus further reducing the incentive of banks to provide loans at competitive rates and contributing to a (temporary) drop in credit expansion of banks.

After stock prices began to decline sharply, it became more and more difficult to raise funds in the capital market. This caused a decrease of the contribution of bonds, even dropping to a negative growth rate in 1993. While the capital market was severely hit by the strong decline of asset prices, bank loans experienced a short-lived renaissance during 1992-1994. Since the benefit of monitored debt is a reduction in agency costs, especially in times of high risk, and a decreasing value of collateral, for many firms the benefits from monitored debt seem to have sufficiently outweighed its cost. Hypothesizing that the mix of monitored debt and arm's-length debt depends on the level of these potential costs, the development of the financial mix structure in favor of bank debt seems plausible during these years. However, after 1994 banks could not further contribute to the positive growth rate of firm debt. Next to the once-again increasing attractiveness of arm's-length finance, loan losses, especially of small- and medium-sized firms[27], and the steep decrease in asset values ate significantly into the equity of the banking sector. After banks came to fully realize the magnitude of the non-performing loan crisis during the mid-1990s, the attractiveness of bank-loans came once again into question. All in all, the importance of bank loans rather declined during most of the second half of the 1990s.

[27] The liabilities of bankrupt firms as per cent of total funding in the corporate sector jumped from 0.18 per cent in early 1990 to 1.53 per cent in the third quarter of 1991 (Bank of Japan 1996).

2.5 Concluding Remarks

Summing up, in the light of aggregate level observation, we can state that large manufacturing firms have enjoyed increasing flexibility in the choice of their capital structure ever since the 1980s. As we will further explore in chapter 4, having (finally) a large number of alternative sources of finance available, these firms use the opportunity to adjust their financial structure to their current needs. Figure 2-2 and table 2-1 in particular highlight the importance of bond financing. The development of the firms' debt structure shows that corporate bonds became a stable source of finance, though not an ever-increasing one. Though bonds did not continue to contribute to the development of the firms' debt structure as they used to during the boom period, with the exception of 1993, bonds could at least manage to keep their importance in the mix of financial sources during most of the 1990s. This once again highlights the general improvement of the availability of arm's-length finance and the increasing competition between arm's-length and intermediary finance due to liberalization.

Next to liberalization and deregulation, the development of Japan's economy in general and asset prices in particular, which are likely to be the largest source of unexpected movements in gross returns, obviously had a substantial effect on the firms' financial position, as well as their choice about the debt mix structure. Accounting for the accelerator effect, the importance of an unanticipated rise in asset prices is even more important for the firms' financial position since it raises net worth more than proportionally. This stimulates investment and, in turn, raises asset prices and with it the market value of firms even further. It furthermore helps to improve the firms' debt to equity ratios and reduces the dependency of bank financed debt (see also chapter 4).

However, even though financial liberalization paved the way for access to many different sources of funds, the development shows that given the current framework of the Japanese financial markets, borrowers can by no means perfectly substitute for intermediary (bank) finance, not even in times when risk is low and growth rates of uninformed finance is exceptionally high. Hence, following theories of financial intermediation, it can be argued that in a market of asymmetric information like that in Japan, intermediaries (still) provide a valuable service in facilitating financing for high-yielding projects that would otherwise not materialize to the same extent.

The most crucial elements of the liberalization process of Japan's financial markets as well as of the boom-bust cycle, which all strongly affected the role of intermediaries during the last two decades, have been highlighted in this chapter. However, so far we have only provided a description of the overall framework in which the changes of corporate finance and subsequently the formation of a new position of banks have been taking place. Many issues concerning the current role of Japanese bank-firm relationships remain unclear. In particular, the functions of banks in a relationship with firms, which go much beyond the provision of capital, have not been addressed. In what way has the central position of banks in

Japanese bank-firm relationships been challenged during the course of financial liberalization and the development of the boom-bust cycle? What are the effects of the banks' monitoring function in an economy where arm's-length finance is increasingly available but a market of corporate control has not yet developed? Was the reaction of bank financing to the demand and supply shocks homogenous to all large firms and if not, did bank-firm relationships have an impact on the demand and supply patterns of debt? Can the weakening of firm balance sheets, as well as future prospects, be compensated by strong relationships to banks? Do bank-firm relationships still help the firm to have easy access to capital and if so, what does this mechanism look like? Do we have to differentiate between different forms of bank-firm relationships and if so, what are the effects of the various kinds of relationships on input factors as well as output factors of a firm?

Given the various degrees of possible bank-firm relationships and the nature of balance sheet factors, the best way to search for answers to these and other questions is an analysis at the level of micro-data rather than aggregate time series alone. To account for the role of bank-firm relationships during the recent cycle, in the next chapters we will provide various analyses of the role of relationships between banks and firms by using firm-level panel data.

Chapter 3

Data Description and Measures of Bank-Firm Relationships

3.1 The Data Base

For the empirical analysis of this study, we make use of a wide range of publicly available quantitative as well as qualitative information. Most of the data originates from standardized, unconsolidated financial statements, as well as supplementary information, all published in Japanese securities reports (*yûka shôken hôkokushô*). These securities reports, which are the most standardized and detailed source of Japanese company data (Göseke 1996), are mandatory for joint stock corporations (*kabushiki gaisha*) listed on the Japanese stock exchange and follow the standards laid out by the Japanese Securities Exchange Law (*shôken torihiki hô*). They are submitted to the stock exchange and the Ministry of Finance, who publishes them after examination. All firms included in the analysis thus belong to the legal form of the joined stock corporation (*kabushiki gaisha)*.

In addition to information originating from *yûka shôken hôkokushô*, we utilize data from the *Japan Company Handbook*, the *kigyô keiretsu sôran* and the *kabuka sôran*[28] of Japanese manufacturing firms listed at the First Section of the Tôkyô Stock Exchange (TSE). *Japan Company Handbook* data includes the ownership structure of firms as well as an industry classification scheme. Information provided by the *kigyô keiretsu sôran* helps us to determine the composition of the board of directors, including the name of the firm or financial institution from which a director has been dispatched. It furthermore discloses the precise source of bank loans and the affiliation with a *kinyû keiretsu*[29]. Concerning capital market data, information originating from the *kabuka sôran* enables us to compile monthly data on the performance of stocks as well as of dividends paid[30].

---

[28] The *Japan Company Handbook*, the *kigyô keiretsu soran* and the *kabuka sôran* are compendia of basic data on all listed firms in Japan published by Tôyô Keizai Inc. They are standard reference works that are widely used in the securities industry.

[29] For a detailed discussion on the six large industrial groups formed around a main bank (*kinyû keiretsu*) see chapter 3.2.4.3.

[30] The amount of dividends paid can also be drawn from the *yûka shôken hôkokushô*. However, the *yûka shôken hôkokushô* only reports the stock price at market value as of the financial statement's annual cut-off date.

Moreover, macroeconomic data like the annual rate of change of Japan's gross domestic product originates from the Bank of Japan's "Economic and Financial DATA on CD-ROM 1999", TOPPAN PRINTING Co., LTD.

The data originating from the various sources of information outlined above have been compiled into a database under the direction of Horst Albach and with the support of the German Research Council (DFG). This database, called the KAISHA database, can be accessed at the Berlin Science Center (WZB). It contains data for 104 firms, all of which were listed on the TSE for the period from 1970 to 1998 (Albach et. al. 1997). Due to the enormous significance of larger firms to Japan's real net output, the database sample comprises manufacturing firms that were among the largest firms listed at the TSE in 1992. It is important to point out that we had the privilege to adapt and expand the original database to suit the needs of our particular research interest. Hence, we have been able to conduct research using data that no source could offer for our analysis until now.

A number of firms included in the original data set were, for various reasons, excluded from our analysis. Some firms were purchased or underwent a merger during the period observed. In order to secure a complete time series for each sample firm and eventually obtain the advantages of an analysis that uses a balanced rather than unbalanced panel data set (see chapter 4.3.2), we had to exclude these firms from our sample. Furthermore, as to financial statements of firms that are subsidiaries of other firms, Coenenberg (1991) and Pellens (1989) note that their informational value is limited and not always valid. Harhoff, Stahl and Woywode (1998) point out that in Germany, liability of subsidiaries is typically passed on to the parent firm in the case of financial distress (insolvency). Harhoff and Körting (1998) show that the characteristics of the subsidiary firm may therefore carry no information about its creditworthiness. We have found our Japanese sample to be similar. To ensure that subsidiary data does not mislead our analysis, we excluded firms in which other enterprises held fifty per cent or more of the shares. This leaves us with a subset of 92 firms (see appendix VI).

To check whether our sample is representative of large Japanese manufacturing firms, we compared the balance sheet information of our sample firms with that of larger samples (e.g. BOJ 1996, p. 86; MOF yearly data). We find that the overall balance-sheet composition as well as its development over the years is very similar (see appendix III for the development of balance sheet items). This underscores that our sample reflects the population of large Japanese manufacturing firms very well.

Following the classification of industries in accordance with the *Japan Company Handbook*[31] (JCH) of 1995/1, the subset comprises five industries within the manufacturing sector: chemicals, electrical machinery, machinery, pharmaceuticals and transport equipment. Hence, with regards to industrial

---

31 The Japan Company Handbook is published quarterly by Tôyô Keizai Inc. in two issues: "First Section" and "Second Section Including OTC-Companies".

coverage, our sample deliberately encompasses a variety of important industries in the Japanese manufacturing sector with many internationally well known firms like Sony, Sharp, Kubota and Kao. The industrial composition of our sample with respect to firm size (in terms of total assets) and number of employees is shown in tables 3-1 and 3-2.

In the sample, size distributions tend to be heavily skewed. Using mean values of the sample to indicate size is hence not particularly informative, and the median value is more relevant. Given our sampling design, the firms are quite large, with median total assets of 226 billion yen and median employment of 3528 employees. Note that firms in the transportation equipment industry are typically much larger than firms in other industries; this is a consequence of the nature of their business. Appendix VI provides a complete list of firms included in our analysis, as well as the securities code number and industry affiliation of each firm.

For our analysis, it is important to note that in comparison with small- and medium-sized firms, each firm included in the sample most likely has easy access to the capital market due to its legal form as well as its size and listing at Tôkyô's stock exchange. Unlike smaller firms that are not listed at a stock market, the firms included in the sample are much more able to use the full variety of outside financial sources made available through financial liberalization.

Our research focuses on evidence from recent developments in the Japanese financial markets during the boom-bust cycle and financial deregulation, as well as on various consequences of these developments for Japanese firms. Our analysis of the sample firms therefore focuses on the years from 1985 to 1998, which allows us a total of 1288 firm-year observations. These fourteen years cover

Table 3-1: Firm size by total assets (in billion Yen)

INDUSTRY	NO. OF FIRMS	OBS.	MIN	MEAN	MEDIAN	MAX	STD. DEV.
Chemicals	27	378	22	323	246	1390	244
Electrical Machinery	25	350	14	1010	247	4470	1240
Machinery	26	364	11	197	108	976	218
Pharma-ceuticals	11	154	80	343	296	944	189
Transport Equipment	3	42	878	1780	1110	4010	1080
Total	92	1288	11	523	226	4470	807

Source: KAISHA database, authors' calculation.

Note: Industry classification in accordance with the Japan Company Handbook of 1995.

Table 3-2: Firm size by number of employees

INDUSTRY	NO. OF FIRMS	OBS.	MIN	MEAN	MEDIAN	MAX	STD. DEV.
Chemicals	27	378	684	3806	3102	13877	2153
Electrical Machinery	25	350	978	18851	6994	81488	22501
Machinery	26	364	391	3390	1583	16725	4044
Pharma-ceuticals	11	154	176 5	5080	4421	11137	2189
Transport Equipment	3	42	135 74	26507	17398	51717	13406
Total	92	1288	391	14301	3528	81488	14301

Source: KAISHA database, authors' calculation

Note: The *yûka shôken hôkokushô* discloses only data on core employees known as *sha-in* (company members), that is permanent or long-term employees who 'entered the firm' (*nyusha* or *shushoku*) upon graduation from school, although a few hired later could also be included. This group includes both, blue-collar and white-collar workers. Therefore, part-timers, temporary employees, *shokutaku* (non-regular staff members) and 'outside' workers (those hired and dispatched by a subcontracting company) are not included.

the entire boom-bust cycle, starting near the beginning of the bubble period, continuing with the period that represents the changes during and right after the burst of the bubble, and concluding with data that reflects recent developments.

Note that we had to overcome a number of obstacles in order to have a complete time series for our analysis. In Japan, the fiscal year starts on April 1 and ends on March 31 of the following year. Of our 92 sample firms, most had chosen the cut-off date for their rendering of accounts to coincide with the Japanese fiscal year. By 1998, only four sample firms had not adopted March 31 as their cut-off date[32]. However, during the period analyzed, a number of firms adjusted the cut-off date for their accounting period to March 31. Hence, accounting periods that are much shorter than the usual one year emerged in the transitional accounting period. Depending on the old cut-off date for the rendering of accounts, the shortened period can range from one to eleven months. In order to create continuous time series, we had to consider that some accounting periods are shorter than one year, and that not all firms have the same cut-off. Since it is most important for the empirical models used in our analysis to have continuous time series, "period comparability" is given priority over "period continuity" when

[32] Nippon Kayaku (4272), OSG Mfg. (6136), Showa Denko (4004), and Toagosei Chemical Industry (4045).

adjusting for structural changes. In other words, every firm should be represented in each "year"[33] with accounts comprising a period of twelve months, rather than avoiding changes in the continuity of the accounts. Discontinuity chiefly applies to inventory at the end of the accounting period, which does not always match the inventory at the beginning of the next period (see Albach et al. 1997 for further details).

Furthermore, during our observation period between 1985 and 1998, the reform of regulations for annual accounts (*zaimu shohyô kisoku*) in 1987 caused changes to "distribution and administrative expenses" in the financial statements. To briefly describe the effect of this reform, after 1987 the number of positions comprising the statements of "distribution and administrative expenses" declined substantially, leaving only a summary of previously more detailed information. We took this change into account in order to secure consistent data that allows "period comparability" (see Albach et al. 1997).

Definitions and detailed descriptions of the variables used in this study are presented in the appendix (see appendix I for the definition of regression variables; appendix II for descriptive statistics of these variables).

Finally, and most importantly for our analysis, it is necessary to introduce variables accounting for the different forms of bank-firm relationships, which the exceptionally detailed data set allows us to capture. Since describing the relationship between banks and firms is obviously a complex matter and is of central interest to our study, the following section deals exclusively with the explanatory variables used throughout our study to comprise a variety of bank-firm relationships. It discusses these measures in connection with different aspects of bank-firm relationships, monitoring, and corporate governance of the Japanese firm.

3.2 Measures of Bank-Firm Relationships

3.2.1 Introduction

In order to analyze the role of Japanese bank-firm relationships, it is necessary to determine criteria that capture their degree of breadth and intensity. Similar to *kinyû keiretsu* affiliations, general bank affiliations were rather stable during the past decades. However, in contrast to *keiretsu* membership, which has

---

[33] The division of business transactions in comparative periods does not follow the calendar year but the time period from September 1 to August 31 of the following year.

continued to be stable even in recent years[34], the dramatic changes of Japanese financial markets and the recent instability of the Japanese economy make it necessary for our analysis to allow for the increasingly dynamic nature of general bank affiliations (e.g. BOJ 1992, Kobayashi et al. 1993, Horiuchi 1994). As will be shown below, the breadth as well as intensity has been continuously subject to change at least since the 1980s.

The rest of this chapter is organized as follows: First, the common interests of banks in bank-firm relationships will be discussed. The chapter continues by underlining the variety of affiliations between large Japanese firms of the manufacturing industry and their banks. The importance of distinguishing between different aspects of bank-affiliations is particularly stressed. The chapter concludes by introducing four measures whose development and interactions are tracked from the mid-1980s. These measures will be used throughout the analysis of different aspects of bank-firm relationships, which will be the main focus of the following chapters.

3.2.2 Common Interests of Banks in Bank-Firm Relationships

Research on Japanese bank-firm relationships is usually limited to analyses of relationships between firms and their main banks (see Aoki and Patrick 1994, among others), or even more limited to main bank affiliations of *keiretsu* firms (e.g. Hoshi, Kashyap and Scharfstein 1990, 1991). However, large manufacturing firms usually have relationships with more than one main bank, several core banks and an additional number of other banks. Depending on the criteria used to determine relationships, a firm can easily have a relationship with ten, twenty or even more banks, the intensity of which can of course widely vary. In contrast to existing research on the relationship between firms and their main banks, the following analysis aims to explore Japanese bank-firm relationships on a much broader basis, where multiple bank contacts are possible. To broaden this analysis and identify interests common to banks, we have to discuss the degree to which banks, despite increasing competition with each other, jointly pursue shared goals.

Modern theory of financial intermediation is based on the concept that, relative to the anticipated benefits, it is too costly for outside investors to continuously learn about all aspects of a firm's activities (Diamond 1984; Stiglitz 1985; Gorton and Haubrich 1986; among others). In this environment, banks rise from being simply financial institutions to being delegated monitors. The economic argument which lies behind this view of banking institutions as delegated monitors runs roughly as follows: Holders of marketable corporate

[34] Regarding our sample, there was only one firm, Kyôwa Hakko (4151), which changed its state from being a non-*keiretsu* firm to a *keiretsu* affiliated firm. All other firms remained either independent or affiliated to their particular *keiretsu* throughout the sample period.

securities have few economic incentives to monitor firms in order to control agency problems. This is especially true for shareholders with few shareholdings in a firm, e.g. individuals with a very diversified shareholding portfolio. The expected personal gain of such individual investors is likely to be small even though the total gains from monitoring might be large; this leads to insufficient monitoring since investors have inadequate incentives. However, provided that the gains from monitoring are sufficiently large, a bank can overcome this incentive problem, because it acts as a delegated monitor by aggregating the capital of a large number of individual investors.

In the following it will be assumed that leading banks (main and core banks) can arise as delegated monitors and therefore act as "trendsetters". Their function as "trendsetters" is vital for all providers of capital. However, this role is particularly important within the group of all banks having a stake in a firm's performance, including for banks whose involvement with a firm is rather small. The reason why the role of leading banks as "trendsetters" is particularly valid for other banks is because among all providers of capital, the group of banks is the most homogeneous: the banks have similar problems and dependencies as well as opportunities and goals.

Banks do not only hold the objective and solely functional position of financial intermediaries as described above. Other than being institutions for aggregating the capital of a large number of individuals and providing it to firms for investment, private banks and their shareholders have interests like any other private firm and therefore have to control for risk while continuously extracting sufficiently high returns for their services. While other providers of capital (for example, shareholders) may have interests similar to those of banks, how they pursue these interests may be very different. Moreover, banks may compete with each other, but their strategies and needs may be similar and thus call for cooperation as well. One might assume that in comparison with other providers of capital, banks act much more congruently in order to accomplish their commonly shared goals.

To further understand the role of a leading bank as a "trendsetter" in our model economy, let us consider a multiple agent framework. Suppose that each of n banks holds an initial endowment of I/n units of investment capital, where $n \geq 2$. In this setting, a duplication of effort occurs because each of the n banks must monitor the firm if a control mechanism is to exist. As in Diamond (1984), a delegated monitor makes an effort to avoid this potential duplication of effort. Though Diamond established a broader theory of financial intermediaries as delegated monitors, where investors in general delegate the monitoring function to financial intermediaries, his theory can also be applied to the more specific setting observed in our analysis. In fact, offering the same products in a similar way, the interests and incentives of the monitoring bank are even more congruous with other banks than with the interests of non-bank providers of capital. Though the magnitude may differ depending on the degree of involvement with a particular firm, all banks face similar problems caused by the need for verifications. This was especially true in Japan until the beginning of the 1980s, when financial markets were strongly regulated and competition among banks was limited. But it still holds true today, in a period where financial liberalization triggers increasing competition, particularly between intermediated and direct finance. Overall, the uniformity of bank interests is much stronger than the divergence caused by competition among banks. For example, as we have seen in chapter 2, in the wake of financial liberalization, bonds, among other products, become increasingly attractive as an alternative and cheap source of capital and therefore are a serious threat to intermediary finance. In this respect, it is rational for the other banks to behave similarly to the leading banks that watch the client firm more closely. Hence, in this sense it is possible to talk about "the common interest of banks". This leads to the ability or even need to analyze general bank-firm relationships by aggregating banks, when possible, with respect to different types of relationships they may have with a firm.

3.2.3 The Need to Distinguish between Different Aspects of Bank-Firm Relationships

The reasons why we picked Japan rather than Germany or the United States for empirically investigating the effects of bank-firm relationships are manifold. One reason is the exceptional importance of such relationships in post-war Japan; not least because of their role, the Japanese system of corporate finance is considered to be bank-centered. Even more important in our view, the relaxation of financing constraints on large Japanese firms during the period under observation provides a natural testing ground for hypotheses on the changing nature and importance of bank-firm relationships when confronted with liberalization. Finally, we picked Japan as the subject of our study because the data needed for the following in-depth analysis are more readily available for Japanese firms than those of other industrialized economies.

Though the availability of data is a helpful advantage, determining the breadth and intensity of bank-firm relationships is not a trivial exercise. There is no commonly shared definition of bank-firm relationships, neither in academic literature nor among bankers or representatives of client firms (Miarka and Yang 1997). For Germany, Albach et al. (1996), assume that there is a bank-firm relationship if one of the following holds: (1) the bank holds more than 10 per cent of the firm's equity; (2) an insurance company holds more than 10 per cent of the firm's equity; (3) the chairman of the firm's board is a member of a bank; (4) the chairman of the firm's board is a member of an insurance company; (5) a member of a bank executive committee is a member of the board and no other bank is represented on the firm's board.

Japanese bank-firm relations are often characterized by cross-shareholdings of stocks, the exchange of personnel, a strong commitment of the bank as a lender of last resort, a high number of bank-loans, and the provision of information between firms and banks (Nakatani 1983, 1984; Aoki 1987; Aoki and Patrick 1994), therefore extending far beyond the capital access function. However, there are many different criteria that have been used in the literature to ascertain the degree and nature of Japanese bank-firm relationships.

To give some examples, a simple measure often used to distinguish between strongly bank-related firms and firms with weak bank relations is to ascertain firms with and without a *kinyû keiretsu* affiliation (see Hoshi, Kashyap and Scharfstein 1991). Unfortunately, this criterion does not consider the fact that it is not necessary to establish a *keiretsu* affiliation in order to have a strong bank-firm relationship. The exclusive use of this criterion is therefore not appropriate for determining a bank affiliation in our analysis. Another method widely used to detect bank-affiliation is to adopt a criterion indicating main bank clients. Classification schemes that utilize a definition of main bank clients are usually based on works like *kigyô keiretsu sôran* (hereafter kks), Dodwell Marketing Consultants' *Industrial Groupings in Japan* or *keiretsu no kenkyû.* Generally, the minor differences between their definitions do not qualitatively affect the results of research on main bank relations (Weinstein and Yafeh 1998).

One of the central elements that all of these definitions have in common is that they identify a main bank by its high degree of loan provision to the individual client firm. A main bank affiliation certainly reveals that the bank has substantial monitoring power, since in addition to the provision of a considerable number of loans, it usually conducts the cash management for its customer and holds an amount of client firm shares that is close to the permissible maximum of five per cent (Art. 65 Securities and Exchange Law; Art. 11 I Anti-Monopoly-Law). Nevertheless, this criterion does not consider the fact that a close relationship with a bank does not necessarily imply a continuously high degree of bank financing. In addition, it does not capture affiliations with other core banks. If anything, these definitions only consider other monitoring instruments such as bank-dispatched directors serving on the client firm's board of directors, or voting power in shareholders' meetings due to bank-owned shares by aggregating these different means which indicate a relationship. Similar to the measure that uses a *kinyû keiretsu* affiliation, the main bank criterion can only be used as a dummy variable, which precludes measurement of the intensity of the relationship. It is therefore not possible to capture the variety of relationships using a single variable. As a result, the use of a criterion provided by any of the works of reference mentioned above might be sufficient to analyze main bank affiliations. However, it certainly has a number of drawbacks when the overall effects of bank-firm relationships are analyzed over a period of time.

As we have pointed out above, bank affiliations have become increasingly dynamic. At least since the 1980's, the intensity and nature of relationships in many cases have been subject to change. Nevertheless, most of the analyses eventually identify the firm's bank affiliation by picking a single date, and recording whether a relationship existed on that date. This obstructs analysis of the development of bank-firm relationships over time. Though relationships had been rather stable in previous decades, since the 1980's, the degree and nature of relationships have been subject to change in many ways (e.g. BOJ 1992, Kobayashi et al. 1993). To allow for this in our analysis, it is necessary to continuously follow the development of such relationships over the period under observation. Hence, it is important to use criteria that capture both various forms and degrees of bank-affiliation.

3.2.4 Comprising Different Ways of Establishing a Bank-Firm Relationship

For our analysis it is essential to develop measures that enable us to incorporate a wide variety of available indicators of bank-firm relationships. In order to take account of transformations of bank-affiliations that may have been triggered by changing economic, legal, or firm conditions, it is furthermore necessary to capture possible changes of these measures over the period analyzed. Hence, we have developed four measures, to be discussed extensively in the following, which incorporate different observable aspects of bank-firm relationships.

Observations were made at four-year intervals (1984, 1988, 1992, 1996). We use data from each of these years to determine the breadth and intensity of a bank-firm relationship for the following four years (or the following two years for the 1996 sample), since their impact cannot be measured sooner than for the period that follows respectively (e.g. using annual data, it is plausible to us that the effect of a bank officer who has been dispatched to a client firm's board of directors in year t cannot be measured sooner than in year $t+1$). The data offers us a way to detect possible changes over a period spanning from the beginning of Japans' financial market liberalization in the early 1980s up to Japan's current crisis in the financial sector. In this way we are able to empirically analyze the dynamic nature of the relationships.

3.2.4.1 Personnel Relationships between Banks and Client Firms at Board Level

Concerning bank executives on corporate boards, two broad economic hypotheses can be identified. One focuses on the Japanese managerial labor market, which views bank executive placements as a reflection of Japan's mid- to late-career managerial labor market. The other argues that bank-dispatched directors maintain ties between banks and their client firm, which facilitates information gathering and monitoring by the bank, as well as the bank's role in corporate governance. Our analysis concentrates on the latter view, though Sheard (1994) points out that these two economic functions are quite complementary. Both emphasize the role of banks in capital market oversight and governance.

Japanese boards – a combined model Since the revision of the Japanese commercial code (*shôhô*) in 1950, the Japanese board system is very similar to the Anglo-Saxon model. Nevertheless, as a remnant of the German-type board system that was in effect in Japan from the 1880s until 1950, it has continued to maintain an additional body of auditors (*kansayaku*), resembling the German *Aufsichtsrat*. When characterizing the Japanese-style corporate structure, we can therefore call it a "combined model" (*setchû-gata heiretsu kikô)* (Yamaura 1990: 44, 1994: 21). It has to be pointed out, though, that the de facto power of the auditors' influence is much smaller than that of their German counterpart.

Japanese corporate boards comprise the top executive officers of the firm, and the majority of directors are internally promoted managerial employees. However, especially large firms also have outside directors[35] that come mainly from large shareholders and banks. To establish a bank-firm relationship based on personnel dispatch at the board level, banks may send one or more of their officers *(haken yakuin)* to the board of directors as well as to the *kansayaku* board. To be considered as a bank-dispatched director in our analysis, the board member must have been with the particular bank at least during most of his career before dispatch. In many cases, he is still on the payroll of the bank while serving the client firm. Having representatives on either of the boards of the client firm provides banks with easy access to information about the firm, whereby they can influence the decisions of the firm much more directly than would be possible from outside. This certainly increases their capacity for monitoring and corporate governance and can diminish the possible moral hazard problem that might accompany a credit relationship. For example, due to the banks' position in the board, they may restrain the entrepreneurs from opportunistically taking a business strategy with a low probability of success (Albach 1999b).

The degree of influence and access to information depends heavily on the number of directors dispatched, as well as on their rank within the "combined model". Since the legal definition and the de facto hierarchy of the boards differ to a high degree, there are many ways to describe the structure within the "combined model". However, for the purpose of our analysis, we focus on the de facto hierarchy that can be observed for the majority of large Japanese firms and is in fact apparent for all our sample firms.

The de facto hierarchy Starting from the bottom of the hierarchical pyramid, the competence of members of the *kansayaku* board is only limited to the right of controlling the board of directors; thus *kansayaku* members have access to first-hand information but no direct influence on the decision-making process of

[35] Note that other than in the U.S. literature, where an outside director refers to a part-time director who typically has multiple board memberships (Black, 1992: 841), we follow the definition used by Sheard (1994: 89) where "outside" refers to the background of the director, rather than his/her current status. In 1991, three quarters of all directors belonging the 2086 listed firms were outside directors and only 5.3 per cent of all directors entered firms from banks (*kigyô keiretsu soran* 1991) .

A measure for bank-dispatched directors Now that we have a general understanding of the Japanese-style structure of the board system, let us introduce an index that considers the composition of the "combined model" and its implications concerning personnel relationships between banks and client firms. The data needed to create this index have been compiled from the *kigyô keiretsu sôran* and cross-checked with information of the *yûka shôken hôkokushô*. The index, which we have called "DISPATCH", has been developed in two steps. First we computed the weighted sum of all board members for each firm and date and the weighted sum of board-executives dispatched from banks. Reflecting the different de facto rankings within the Japanese-style structure of the board system, which determine the level of influence on the decision-making process as well as the degree of access to information, the weights used are as follows:

Auditor (*kansayaku*): 1 point,

Ordinary board member: 2 points,

Member of the representative board (*jômukai*): 5 points.

In a second step, the portion of board-executives dispatched from banks has been calculated out of the weighted sums computed in the first step. As a result, a comprehensive index is available for our analysis, which expresses the personnel ties between banks and firms at board level and which considers the different position of bank-dispatched board-members, as well as the varying size of boards which usually differ over time and by firm.

The development of DISPATCH As can be seen in tables 3-3 and 3-4, the mean of the overall index declines from the beginning of the observation period and only slightly increases in 1996. However, it does not once again reach the level of the first half of the observation period. Also, the number of firms with no personnel relationships with banks increases continuously until it decreases sharply in 1996 (table 3-4). This demonstrates that personnel ties were reduced over the bubble period and then increased again slightly during the depression that Japan has to deal with ever since the bubble burst, therefore supporting what has already been tentatively suggested above.

Table 3-3: Descriptive statistics of DISPATCH

YEAR	N	MEAN	25% PERCENTILE	75% PERCENTILE
1984	92	5.414	0.000	7.634
1988	92	4.847	0.000	7.968
1992	92	4.163	0.000	6.257
1996	92	4.364	0.000	5.815

Source: KAISHA database, authors' calculation

Table 3-4: Development of the DISPATCH-index

DISPATCH Index	1984	1988	1992	1996	1984-1996
0	33 firms	36 firms	37 firms	32 firms	23 firms
0 > x <10	42 firms	41 firms	44 firms	45 firms	24 firms
10-	17 firms	15 firms	11 firms	15 firms	7 firms
Index-Sum	498	446	383	402	-

Source: KAISHA database, authors' calculation

Even though personnel relationships at board level seem to slightly increase once again during times of economic distress, the trend clearly indicates that this form of relationship is overall becoming less common. It can be argued that from the banks' point of view, it was less necessary to oversee and manage the decisions of client firms during the economic boom period, because the firm data available suggested that most firms were doing exceptionally well. However, banks have had to deal with the severe consequences of this misjudgment since the early 1990s. In addition, Ogishima (1994) points out that banks are increasingly focusing on a smaller number of firms, so that in the later years of the observation, firms usually do not have bank representatives from a variety of banks on their board, but only from one or two banks. The more plausible interpretation, however, is that in the wake of financial liberalization, increased access to outside finance other than bank finance has allowed a growing number of firms to loosen their bank ties, thus decreasing the number of bank representatives on the board. This trend has also been indicated by other studies (BOJ 1992; Ogishima 1994). In any case, the use of financial sources other than bank loans was especially common during the bubble period and continues to be valid even in times of a depressed economy. This certainly has a strong effect on bank-firm relationships, and implications of this development have to be closely analyzed.

Though the overall trend of the index is to decrease, we observe a slight (exceptional) increase of the index maximum in 1992. As has been mentioned above, this development can be explained by an anticipation of the last revision of the Commercial Law (*shôhô*). Many firms invited one or more additional board members from a bank to take a position as "outside" auditor. This largely accounts for the slight and short-lived interruption of the overall trend observed. In addition, table 3-4 shows that the number of firms with a continuously large DISPATCH-index of ≥10 points is small. This suggests that only a few firms continuously receive representatives from more than one or two banks.

3.2.4.2 Shareholding of Banks

The influence of shareholders The general shareholders' meeting (*kabunushi sôkai*) is de jure the highest body in a Japanese stock corporation (*kabushiki gaisha*). Here, a strict one-share one-vote rule has to be followed (Commercial Code Art. 241(1)). In the meeting, shareowners elect, in principle, the members of the board of directors as well as of the *kansayaku* board. The Commercial Code furthermore demands the approval of the balance sheet and income statement, and the setting of dividends.

However, it is widely agreed that a shareholders' meeting is more a legal formality (Gerlach 1992: 234) than a powerful body that conducts corporate governance. Insofar as shareholders have influence, the actual decisions are usually formed by a handful of strong and oftentimes interrelated shareholders before the actual meeting takes place (Fukao 1995). These shareholders do little trading and can therefore be considered stable shareholders. In many cases, representatives of this group of shareholders do not even show up at the general shareholders' meeting, thus furthermore underlining the unimportance of this legal event. In other words, like the board of directors satisfies a legal demand by approving decisions made by the *jômukai*, so the general shareholders' meeting provides legal approval for the decisions of pre-decided resolutions passed by a small power group of large and influential shareholders. Also, compared to other (individual) shareholders, this group of large and most often institutional shareholders has much better resources for information gathering and monitoring, in other words, corporate governance. Trying to minimize their agency-costs, smaller shareholders therefore most likely act as free riders.

Bank-held shares Within this group of large and influential shareholders, banks often play a central role. In fact, there have been pronounced trends in the composition of share ownership in favor of institutional investors in general and banks in particular. Since the end of World War II, the percentage of shares held by financial institutions has been increasing more or less continuously, from 9.9 per cent in 1949 to 38.7 per cent in 1997, with banks owning more than half of these shares. Over the same period, individual holdings declined from 69.1 per cent to 24.6 per cent of total shares[42] (*kigyô keiretsu soran* 1999: 81). This trend developed despite Art. 11 I of the Japanese Anti-Monopoly Law (*dokusen kinshi hô*), which allows banks to hold only up to five per cent of a single firm[43] (prior to 1987 up to ten per cent) on their own account. Note that banks are not allowed to

---

42 The percentage of shares held by financial institutions peeked in 1989, reaching 42.5 per cent, while shareholdings of individuals reached an all time low with 22.4 per cent that same year.

43 It has to be noted though that in our sample, the shareholdings of trust banks exceed this five per cent level in a number of cases even after 1987. This is due to fact that shareholdings of trust banks include stock investment for the purpose of investment trust and pension trust management.

use proxy-votes as is possible for banks in Germany. Among the five top shareholders of any of our sample firms, we usually find at least one bank, a figure that corroborates other works of research (Aoki, Patrick, and Sheard 1994). Although competition between banks is increasing along with financial liberalization, banks usually manage to coordinate themselves in cases where their interests (for example, their common interests as lenders) have to be pushed through against the interests of other shareholders (Jensen and Meckling 1976; Smith and Warner 1979; Barnea, Haugen and Senbet 1985). Considering the fact that banks own on average more than 22 per cent of the shares of a sample firm (table 3-5), it is clear that the combined effect makes it very difficult to act against their collective interests. Besides jointly establishing their position as debt-holders, large equity holdings help banks to establish an obligation that cements a relationship with a firm and creates captive customers (Beason and James 1999). In addition, banks are among the most stable shareholders, which furthermore consolidates their position.

Hence, in addition to dispatched directors, bank-owned shareholdings of client firms are another powerful tool for banks that facilitates monitoring and secures the bank's role in corporate governance. In order to capture the magnitude of this tool in our analysis, we compute the percentage of client firm-shares held by banks among the 20 largest shareholders. The data has been drawn from the spring edition of each particular *Japan Company Handbook* (JCH) and was cross-checked with information on the stockownership structure provided by the *yûka shôken hôkokushô*. Stock holdings are expressed as a percentage of the total, up to one decimal place.

The development of SHARE To briefly describe the development of this measure, called "SHARE", tables 3-5 and 3-6 show a continuous increase in shares held by banks until 1992 and indicate a small decline for the sample year 1996. The increase during the bubble period can be explained by at least two factors.

Table 3-5: Descriptive statistics of SHARE

YEAR	N	MEAN	25% PERCENTILE	75% PERCENTILE
1984	92	17.586	13.380	22.120
1988	92	22.962	17.215	27.125
1992	92	24.181	19.235	28.240
1996	92	23.476	19.170	28.390

Source: KAISHA database, authors' calculation

Table 3-6: Development of SHARE

PERCENTAGE OF SHARES	1984	1988	1992	1996	1984-1996
<10	9 firms	5 firms	0 firms	3 firms	0 firms
10 ≥ x <25	69 firms	50 firms	48 firms	55 firms	27 firms
≥25	14 firms	37 firms	44 firms	34 firms	12 firms

Source: KAISHA database, authors' calculation

First and most important, banks participated massively in the speculative bubble. Similar to firms outside of the financial industry, banks increased their equity by issuing large amounts of stocks. These stocks were then usually exchanged with client firm shares, thus strongly improving the balance sheets of banks as well as their client firms. Though this procedure was de facto only an exchange of paper (Scher 1997), it deprived the influence of other shareholders, notably private ones.

Second, the micro data of the KAISHA database, which allows us to distinguish between the different types of banks (city, long-term, trust, foreign), indicates an especially large increase in stocks held by trust banks. Note that trust banks are allowed to manage pension trusts and investment trusts, which strongly increased their investment portfolio in favor of stock holdings. The increase thus most likely reflects the participation of these investments in the speculative boom of the Tôkyô stock market.

After the burst of the bubble, banks had to further increase their equity ratio in order to meet Bank of International Settlement (BIS) requirements. Since they were not able to do so by further issuing new stocks, banks slowly began to raise equity by selling stocks of their portfolio. Though in 1992 banks held at least 25 per cent in almost 50 per cent of our sample firms, the overall number of firms in which banks continuously hold such a high stake over all sample years is small (12 firms). This demonstrates that the exceptionally large number of stocks held by banks during the bubble period is partly due to speculation and therefore short-term, rather than for long-term motives. However, we can conclude that banks are among the largest and most stable shareholders over the whole observation period and thus have a strong potential to monitor and influence client firms via their shareholder position.

3.2.4.3 Kinyû Keiretsu Affiliation under the Aspect of Loan-Dependency

The Japanese industrial structure is well known for its large industrial groups (*kigyô shûdan*), widely referred to as "*keiretsu* groups" in the Western literature. Concerning the structure of these groups, one has to distinguish between three kinds of *keiretsu*: The center of a *keiretsu* group is formed by a *kinyû keiretsu,* which is generally understood as an industrial group with a number of large firms around a (city-)bank and a general trading company (*sôgô shôsha*). The large industrial firms that are members of a *keiretsu* are furthermore embedded in distribution networks called *ryûtsû keiretsu* and production networks called *seisan keiretsu* (see Gerlach 1992; Lincoln, Gerlach, and Takahashi 1992; Steinbrenner 1997; Dolles 1997; Moerke 1999, for an in-depth discussion of *keiretsu*).

Defining kinyû keiretsu affiliation Of these three types of *keiretsu*, the *kinyû keiretsu* is the one which is mainly based on financial aspects (provision of loans, mutual shareholdings, etc.) and opportunities to exchange information, rather than the transaction of goods. Hence, for the purpose of our study we focus on *kinyû keiretsu* affiliations.

However, there are a large variety of possible *kinyû keiretsu* structures. Of the numerous definitions which can be found in the literature, we follow Ito and Hoshi (1992), who claim that a firm belongs to a *kinyû keiretsu* group when it is a regular member of a "Presidents' Council" (*shachôkai*)[44]. Since there are six different *shachôkai,* we can differentiate between six large horizontal business groups, usually addressed as *roku daikigyô shûdan*. By conservative estimates, 89 of the top 200 Japanese industrial firms are associated with one of these six horizontal groups (Hoshi, Kashyap and Scharfstein 1991: 37). In our sample, 33 out of 92 sample firms belong to one of these groups (see table 3-7).

Using the criterion of *shachôkai* membership has many advantages. Due to the yearly publication of the *shachôkai* member lists by Tôyô Keizai's *kigyô keiretsu soran,* we have a reliable and continuously updated source that identifies to which, if any, *shachôkai* each individual sample firm belongs[45]. Furthermore, this definition has already been used by a great number of studies, including Moerke (1999), who not only used the same definition but also successfully used the same sample of firms for his study on performance and corporate governance of *keiretsu* groups. Moreover, following this definition, the sample distribution

---

44 The *shachôkai* regularly hold meetings where the top CEOs of firms belonging to the particular *kinyû keiretsu* exchange information. Though there is no evidence that strategic decisions are made during these meetings (Kôsei torihiki iinkai 1992), it is widely perceived that this institutionalized exchange of information is the key signal to identify the members of each particular *kinyû keiretsu* group (Ito and Hoshi 1992; Moerke 1999).

45 In a small number of cases, firms are members of more than one *shachôkai* at the same time. In these few cases, we assign the firm to the *keiretsu* that has, according to the degree of shareholdings and loan provision, the strongest influence on the firm.

Table 3-7: Distribution of sample firms by *kinyû keiretsu* and presidents' council

KINYÛ KEIRETSU	PRESIDENTS' COUNCIL	NUMBER OF FIRMS
Mitsui	Nimoku kai	2
Mitsubishi	Kinyô kai	5
Sumitomo	Hakusui kai	4
Fuyô	Fuyô kai	4
Sanwa	Sansui kai	6
Daiichi*	Sankin kai	12
No affiliation	No affiliation	59
Total		92

Note: *kinyû keiretsu* and Presidents' Council classification in accordance with the *kigyô keiretsu soran*, several issues. *Kyôwa Hakko Kogyo Co. (4151) is member of the Daiichi group since 1988.

remains relatively constant over the years. In fact, not a single sample firm changed its *shachôkai* membership and only one unaffiliated firm (Kyôwa Hakko) became member of a Presidents' Council during the fourteen years under observation.

Besides characteristics like interrelating shareholdings and an exchange of personnel, one of the often claimed advantages of a *kinyû keiretsu* affiliation that is of interest to us is the easy access to loans provided by other *keiretsu* members, especially those belonging to the financial sector. In fact, among the various links between *keiretsu* firms (e.g. between the general trading house and manufacturing firms, or between the manufacturing firms themselves) the most important financial link is between *keiretsu* firms and the main bank at the center of each industrial group. Note, however, that the already-mentioned city bank, which most often takes the position of the main bank of all group members (see also 3.2.4.4), is not the only financial institution which belongs to a particular *kinyû keiretsu*. An industrial group comprises additional firms from the financial sector, like other banks (most likely trust banks) or insurance companies. Also, during the years of the bubble economy, many banks, including main banks, established financial institutions that provided firms with massive amounts of additional loans. These institutions were legally separate from banks but de facto controlled by them. In contrast to *keiretsu*-independent firms that are much more inclined to spread their borrowings, *keiretsu* firms usually do a significant fraction of their borrowing from the banks of their own group. Together with large equity stakes held by the banks of the same *keiretsu*, this concentration of financial claims in the hand of a few banks is likely to mitigate the free-rider problems that firms with diffusely held debt and equity have to deal with (Hoshi, Kashyap and Scharfstein 1991).

We also have to take into account that, when it comes to the provision of *keiretsu* loans by financial institutions, the main bank takes a central position. Besides the loans that a main bank itself supplies, it coordinates the loans that other *kinyû keiretsu* members allocate to a firm as well. Holding this position, the main bank therefore has access to a wide range of information. Its capacities to influence a firm via loan provision go much beyond the influence it would gain solely on the basis of its own loan provision. Hence, due to the leverage a main bank of a *keiretsu* may have, *kinyû keiretsu* affiliations are certainly another fundamental aspect besides bank shareholdings and board-level personnel relationships that has to be included when analyzing bank-firm relationships.

However, when constructing a measure of *kinyû keiretsu* affiliations for our analysis, we have to consider recent developments concerning *kinyû keiretsu* affiliations, especially with respect to loan-dependency, which may influence the leverage of main banks of *keiretsu* firms. Since the mid-1980s, being member of a *keiretsu* does not necessarily go along with a high degree of *keiretsu* loans (Abrahams and Hess 1998). Though characteristics of *keiretsu* firms like bank shareholdings and bank-dispatched directors remain rather stable, with financial liberalization firms have tended to reduce their demand for loans originating from financial institutions of their *keiretsu* and rather have used other sources of outside capital. In other cases, especially after the burst of the bubble, weakened financial institutions have circumvented the provision of loans for ill-performing member firms of their *keiretsu* group, since loan provision for these firms is perceived to be too risky. Both factors certainly have influenced the leverage of *keiretsu* banks mentioned above.

A measure for *keiretsu* bank-firm relationship Considering the general position of a *keiretsu* bank as well as the recent developments of *keiretsu* loans described above, a proxy is needed that incorporates the intensity of the *keiretsu* bank-firm relationship especially with respect to the changing leverage of banks which coordinate *keiretsu* loans. One way to create an appropriate proxy would be to measure the amount of finance a firm actually received from financial institutions of its *keiretsu* in contrast to loans from other financial institutions. Unfortunately, individual loan contracts are not publicly accessible. However, in addition to balance sheet information about bank loans, the *yûka shôken hôkokushô* provide information on the amount of loans given to a particular firm per financial institution. We collected this data for each of our client firms and cross-checked this information with similar information provided annually by the *kigyô keiretsu soran*.

As a result, we are able to construct a measure which reflects the intensity of *keiretsu* bank-firm relationship with special emphasis on the aspect of loan-dependency as follows: We divide the sum of loans provided by financial institutions of a firm's *keiretsu* by the total sum of loans all firms in the financial sector have appropriated for the client firm. This measure, which is called KEILO in the following, has been computed yearly for each respective firm.

Table 3-8: Descriptive statistics of KEILO

YEAR	N	MEAN	25% PERCENTILE	75% PERCENTILE
1984	92	7.762	0	16.612
1988	92	7.962	0	15.946
1992	92	7.960	0	15.572
1996	92	7.924	0	16.491

Source: KAISHA database, authors' calculation

Sample results of KEILO The descriptive statistics presented in table 3-8 underscore that, despite strong economic fluctuations during the Japanese boom-bust cycle, the influence of *keiretsu* loans is very stable over all sample years. Hence, our results emphasize that the provision of *keiretsu* loans, and with it the leverage of *keiretsu* main banks does not decline overall. Considering only the *keiretsu* affiliated firms, which make up about one-third of our sample, the mean of KEILO is around 16 per cent. This finding corroborates results of an analysis by the *kôsei torihiki iinkai* (1994), which includes the source of borrowings of all firms affiliated with one of the "big six" industrial groups.

Table 3-9 provides us with a more detailed picture. For more than half of the firms with *keiretsu* affiliation, *keiretsu* loans continuously play an important role[46]. However, in 1984 as well as 1988, we have cases where firms have no *keiretsu* loans at all, despite an affiliation with an industrial group. In addition, more than a quarter of the sample firms with *keiretsu* affiliation show only a small

Table 3-9: Development of KEILO

KEILO-RATIO	1984	1988	1992	1996	1984-1996
0	61 firms	60 firms	59 firms	59 firms	59 firms
$0 > x \leq 15$	6 firms	9 firms	10 firms	10 firms	1 firm
> 15	25 firms	23 firms	23 firms	23 firms	19 firms

Source: KAISHA database, authors' calculation

46 For some cases, the share of loans provided by financial firms of the same *keiretsu* reaches close to fifty per cent.

number of loans originating from financial institutions of the same industrial group. Note that the number of such firms rises slightly over time, which suggests a gradual decrease of the importance of *keiretsu* loans for certain firms, despite a stable and continuous affiliation with an industrial group.

Nevertheless, we can conclude so far that *keiretsu* loans overall seem to play an important role for firms with a group affiliation throughout the period under investigation. Therefore, the leverage of the groups' main banks continues to have a strong basis, which underlines the importance of considering *keiretsu* affiliations in our study.

3.2.4.4 Main Bank Affiliation

In contrast to the market-oriented, decentralized financial system of Anglo-Saxon economies, we can often observe firms with a main bank and a number of additional core banks in the well-developed system of relationship banking found in Japanese-German type financial markets. Japanese as well as Western scholars (e.g. Nakatani 1984, 1990; Imai 1990; Sheard 1989; Georg 1996) have underscored the advantages of the Japanese form of financial organization, especially highlighting the function of main banks for industrial development. International institutions like the World Bank (see Aoki and Patrick 1994) even claim that the Japanese main bank system may serve as a model that should be emulated by emerging economies of the developing world and transforming economies of the former socialist countries in Europe alike.

On the other hand, even before the current banking crisis, some works have claimed that the function of Japanese main banks is nothing more than a myth (Miwa 1985, 1991, 1996). Taking a more moderate position, the Japanese 1992 Economic White Paper argues that, based on the autonomous decisions of each entity, the relationship between firms and main banks will undergo further change and diversify more rapidly, as reforms are carried out in the Japanese financial system (Keizai kikaku chô 1992).

Obviously, the contentious issue of Japanese main banks attracts great attention and is the subject of a variety of research interests as well as discussions on policy implications. It is therefore crucial for our analysis to include a variable that represents this prominent feature of the Japanese financial system.

Defining Japanese main bank relationships There is no commonly shared definition of a Japanese "main bank" or a "main bank relationship". Trying to capture what most notions of a "main bank relationship" have in common, Scher (1997: 95) states that it is a "relational system of designated rewards and protection based on tradition of insider-outsider status". As will be explained below in more detail, the foundations of this system comprise extensive formal and informal contacts between the firm and its main bank at a variety of levels.

Japanese main bank relationships are usually long lasting relationships that are often, but not necessarily, embedded in a membership of the same *kinyû keiretsu*. In fact, there are many cases where the group bank of a *kinyû keiretsu* that generally serves as the main bank for all members of that particular *keiretsu* is the successor to the prewar *zaibatsu* bank, thus underscoring the oftentimes long tradition of such relationships. These long-lasting relationships may imply benefits for both entities: Main banks build up particular capabilities and expertise to meet the firm's needs on the one hand, and on the other hand secure privileged access to lucrative services.

Advantages for main banks Due to the main bank's expertise as well as a particularly strong commitment to the firm (and vice versa), the main bank is typically predestined to provide many services much beyond supplying outside capital. In fact, as for our sample, the main bank may manage more than 50 per cent of the client's fee-based business, although it usually supplies less than 20 per cent of the client's bank borrowings. With lending diminishing in profitability, services other than lending become the central source of profit for banks (Keizai kikaku chô 1992; Aoki and Patrick 1994; Scher 1997). Main banks regard their position as a particularly lucrative one, being privileged to offer among all competing banks by far the lion's share of these services to the client firm and therefore being able to collect direct or indirect fees for these services over a long period of time (Aoki, Patrick and Sheard 1994; Scher 1997; among others). To give some examples, the need to keep a compensating balance forces the client firm to keep an oftentimes large amount of money in its transaction account. Most often the firm keeps this account, which is an important source of profit for the bank since it is non-interest-bearing, at the main bank. Attending to large, internationally operating client firms like many of those in our sample, the provision of fee-based products like foreign exchange transactions, letters of credit, exchanging foreign currency, swap transactions, underwriting, as well as a large number of newly developed products certainly also belong to the more lucrative services. The client firm is especially likely to turn to its main bank for such services. Furthermore, the main bank is usually entrusted with the personal accounts of employees of the client firm, which creates a solid basis for business in the bank's high-profit-margin retail sector. Note, however, that due to liberalization, arrangements become looser and depend much more upon the needs of each firm and the capabilities of each bank.

Advantages for client firms This relational system can also certainly imply advantages from the firm's point of view, such as designated rewards and protections. Main banks are well known for securing the client firm's easy access to capital. Hoshi, Kashyap and Scharfstein (1990, 1991) show that, indeed, due to an easy access to bank loans, firms with a main bank affiliation are less liquidity-constrained than other firms in their investment decisions, although they do not detect any significant differences in investment levels. Main banks may furthermore provide several insurance functions. There is reported evidence that main banks may act as a lender of last resort if a client firm is in financial distress (e.g. Yabushita 1992). Furthermore, in difficult times a firm may suddenly find several of its top executives replaced by bank personnel, who can provide the firm with managerial know-how (e.g. Aoki and Patrick 1994). The main bank thus leads the rescue or becomes the leader of a settlement (Pascale and Rohlen 1983; Abegglen and Stalk 1985; Sheard 1985). Also, main banks seem to protect their client firms against strong fluctuations of interest rates (Nakatani 1984, among others).

All of these close interactions between the firm and its main bank secure the main bank's access to a large amount of highly relevant, first-hand information. This gives the main bank the incentive and ability to monitor the firm and may reduce information and incentive problems (Hoshi, Kashyap and Scharfstein 1991). It forms the basis for the main bank's strong role in conducting corporate governance and acting as a "lead bank" when it comes to securing the common interests of banks as has been discussed in 3.2.2. Furthermore, it can serve as an implicit guarantee[47] or at least an important signal for the firm's suppliers as well as for outside investors. In fact, this signaling effect may even further secure access to additional sources of capital.

Nevertheless, recent developments suggest that due to the weak position of many banks and the high risks involved in many services that main banks used to offer, main banks are increasingly inclined to circumvent what has been described by many authors as an implicit obligation (Nakatani 1983, 1984; Aoki 1984; Sheard 1989; Aoki and Patrick 1994). It is therefore far from clear whether the stylized facts of a main bank relationship are still valid for the period under investigation.

Developing a main bank dummy We created a dummy variable called "MAIN", taken to be unity if a main bank affiliation exists and zero otherwise. To recognize a sample firm's main bank, we follow the criteria used by the *kigyô keiretsu soran*. According to these criteria, the bank that provides a firm with the largest amount of borrowings among all financial institutions is assumed to be the firm's main bank. This assumption is well chosen since the client firm strictly observes the hierarchy among the lending institutions. Though the distribution of fee-based services is made not only according to the lending position but also depending on the individual capabilities of the bank, there is a tacit understanding that when it comes to lending, the client firm makes sure that for each fiscal year, the bank which has been chosen to serve as the main bank has the lead share in lending (Scher 1997).

Sample results of MAIN Table 3-10 underlines that the number of firms with main banks decreases after reaching a peak during the bubble period. The numbers increase again during the recession, when some firms may have more difficulties taking up capital on the open market and therefore turn back to

---

[47] Taking the view of suppliers and outside lenders, Wallich and Wallich (1976: 273) describe the main bank's function as follows: "It is taken for granted that the main bank assumes a special responsibility with respect to the borrower. In an emergency other creditors can expect their claims to effectively though not legally outrank those of the main bank."

Table 3-10: Descriptive statistics and development of MAIN

MAIN-DUMMY	1984	1988	1992	1996	1984-1996
0	16 firms	14 firms	21 firms	18 firms	9 firms
1	76 firms	78 firms	71 firms	74 firms	64 firms
Mean	0.826	0.848	0.772	0.804	0.814 (all firms, all years)

Source: KAISHA database, authors' calculation

established relations with a main bank. Not surprisingly, more than two-thirds of the sample firms continuously have a main bank during the whole period observed, while less than ten per cent of the firms have no relationship with a main bank during any of the four sample years. After having accounted for bank mergers, out of the 64 firms which continuously have a relationship with a main bank, less than twenty per cent change from one main bank to another.

Note that the overwhelming number of firms which decide to change their main bank belong to a *kinyû keiretsu*, and that the change in lead lending status in a particular year is usually the result of a temporary increase in lending by the trust bank of the same *kinyû keiretsu* for some expenditure that might have been beyond the lending capabilities of the firm's "original" main bank. Most firms return to their former main bank shortly thereafter (Scher 1997, finds similar results for a different sample). The result of an overall stable and long-term relationship is not surprising, since lasting changes from one main bank to another, if at all, are more likely for small- and medium-sized firms. These firms might "graduate" from their local city bank due to overwhelming success and have to look for a new partner who can offer all the services needed for the now large, oftentimes internationally operating firm. Looking at the size of our sample firms (chapter 3.1, tables 3-1 and 3-2), all firms already "graduated" and their main banks are one of the large banks which can offer the full range of financial services.

As has been already pointed out above, relationships between main banks and client firms are oftentimes embedded in a membership in the same *kinyû keiretsu*. As for our sample, only very few firms are a member of a *kinyû keiretsu* without being affiliated with a main bank (table 3-11). In these few cases, firms simply did not ask for any bank loan as of the cut-off date for the particular financial year. The bank which, according to the definition of the *kigyô keiretsu soran*, takes the position of the main bank in the previous and/or following sample year was in all cases entrusted with the personal accounts of the client firm's employees as well as with the client's transaction accounts during the year in which the firm did not indicate a main bank affiliation.

Table 3-11: Statistics of main bank affiliation (MAIN) and *kinyû keiretsu* affiliation (KK)

	1984	1988	1992	1996	1984-1996
MAIN and KK	29 firms	33 firms	32 firms	32 firms	27 firms
MAIN but no KK	47 firms	45 firms	39 firms	42 firms	37 firms
No MAIN no KK	13 firms	14 firms	20 firms	17 firms	8 firms
No MAIN but KK	3 firms	0 firms	1 firm	1 firm	0 firms

Source: KAISHA database, authors' calculation

Though a *kinyû keiretsu* affiliation usually goes along with a main bank relationship, table 3-11 shows that more than two-thirds of the sample firms independent of a *kinyû keiretsu* also maintain main bank affiliations. However, about twenty per cent of all sample firms are independent of both a main bank and a *kinyû keiretsu,* indicating a surprisingly large number of totally independent firms. Owing to increasing financial liberalization, this group of firms grows over most of the sample years, which may be a sign of the decreasing importance of either kind of close affiliation.

Summing up, the results presented in tables 3-9 and 3-10 show some tentative signs that suggest a declining importance of main bank affiliations. Nevertheless, despite financial liberalization and an increasing availability of other financial sources, most firms maintain a relationship with a main bank as defined by the *kigyô keiretsu soran.* Relationship banking in general, and the affiliation with a main bank in particular, still seem to be most common for large Japanese manufacturing firms. However, whether this still implies all the advantages of a main bank affiliation as have been claimed for the period preceding the 1980s has to be analyzed in the following chapters.

3.2.4.5 Interrelation between Different Ways of Bank-Firm Relationships

Critics of the bank-based financial system have pointed out that the different means of bank influence are not independent of each other (Kaserer and Wenger 1998), and that the accumulation of these different means are the bases of the powerful position banks successfully defend. Especially in the case of Japan, as has already been pointed out above, it is widely believed that main bank affiliations are a function of bank-dispatched directors and a large amount of bank finance as well as client firm shares held by banks. In fact, since the general share-holders' meeting decides, for example, on the composition of the board of directors, one can argue that a number of means of bank influence are actually interconnected by the corporate legal structure. On the basis of our sample firms, we will proceed with our analysis by further observing to what degree the different forms of bank-firm relationships are interrelated.

Certainly, when comparing firms with and without a main bank affiliation, banks are an important source of loans only for those firms that are affiliated with a main bank. However, this example alone is not very convincing since it is simply due to the definition of main banks that is usually followed in analyses of main bank affiliation. Nevertheless, results of our sample presented in table 3-12 further corroborate the view that different means of bank influence are not independent of each other: Main bank relationships go along with a high potential for bank influence on the boards of client firms via bank-dispatched directors. While the index of bank-dispatched directors reaches only a mean value of 2.88 for firms without a main bank, the mean value for firms with a main bank affiliation is almost double. In order to test for the statistical significance of our result, we exercised a Oneway-Anova-Analysis, for which we used the dummy variable for main bank affiliation as our grouping variable. We find highly significant differences among the two groups; hence we can reject the H_0 hypothesis of equal means with high statistical certainty (p=0.000).

Table 3-12: Interrelation between main bank affiliation and bank-dispatched directors

	MB	
	0	**1**
Number of observations	69	299
Mean of DISPATCH	2.88	5.17

Source: KAISHA database, authors' calculation; n=368

Table 3-13: Interrelation between main bank affiliation and bank-held shares

	MB	
	0	**1**
Number of observations	69	299
Mean of SHARE	26.00	20.90

Source: KAISHA database, authors' calculation; n=368

Table 3-14: Interrelation between the index of bank-dispatched directors and the percentage of client firm shares held by banks

	DISPATCH		
	0	**< 10**	**≥ 10**
Number of observations	138	172	58
Mean of SHARE	23.11	20.70	22.18

Source: KAISHA database, authors' calculation; n=368

Table 3-15: Interrelation between the percentage of client firm shares held by banks and the degree of *kinyû keiretsu* affiliation under the aspect of loan dependency (KEILO)

	SHARE		
	< 10	**10 ≥ x < 25**	**≥ 25**
Number of observations	17	222	129
Mean of KEILO	23.80	22.96	21.97

Source: KAISHA database, authors' calculation; n=368

In addition, the positive correlation between a main bank affiliation and the degree of *kinyû keiretsu* affiliation under the aspect of loan dependency is also most obvious, since there are almost no *keiretsu*-affiliated firms without a main bank.

However, even though the main bank usually holds the highest amount of client firm shares among all banks (usually close to five per cent, due to legal restrictions mentioned above), our findings do not corroborate that main bank affiliations go along with a high percentage of client firm shares held by the group of banks. In fact, the results presented in table 3-13 even suggest the opposite: As a sum, the percentage of bank-held shares is higher for firms without a main bank affiliation than with a main bank affiliation. The Oneway-Anova-Analysis shows a high level of significance (p=0.0000), which suggests that the difference between the mean value of the two groups is not coincidental.

Looking at the interrelation between bank-dispatched directors and bank-held shares, the results shown in table 3-14 give evidence that not all means of bank-firm relationships are necessarily dependent on each other: Our sample does not come up with a clear pattern that could indicate a straightforward relationship between the degree of influence bank-dispatched directors have on either of the two boards on the one hand, and the amount of client firm shares controlled by banks on the other hand. In fact, showing no evidence of a clear interrelation might even support the claim often made by bankers that bank-dispatched directors are not elected to the boards simply due to the banks' leverage in the shareholders' meetings. At least, there seem to be additional reasons why banks can gain influence via bank-dispatched directors. In following chapters we will consider this issue more closely.

Testing for a correlation between the degree of bank influence via bank-dispatched directors and the degree of *kinyû keiretsu* affiliation under the aspect of loan dependency, the results do not show any clear pattern either.

To our surprise there seems to be a slightly negative relation, if any, between the percentage of client firm shares held by banks and the degree of *kinyû keiretsu* affiliation under the aspect of loan dependency (table 3-15). However, using Oneway-Anova-Analysis, the level of significance for this interrelation (p=0.5565) does not allow us to reject the H_0 hypothesis of equal means. Hence, one should be careful when interpreting the negative relation of these two kinds of bank influence. In any case, simply the fact that there is no clear positive relation between KEILO and SHARE shows that we cannot accept the widely stated view that compared to unaffiliated firms, *keiretsu* affiliation implies a higher percentage of bank-held shares.

Since this result is quite different from previous findings reported in other works of research (e.g. Gerlach 1992), we also tested the relationship between *keiretsu* affiliation and the percentage of bank-held shares of client firms by using the more common method of constructing a dummy variable taken to be unity if a *keiretsu* affiliation exists and zero otherwise. However, the results further corroborate our findings that a *keiretsu* affiliation does not correlate with a higher percentage of shares controlled by banks (see table 3-16).

Table 3-16: Interrelation between *kinyû keiretsu* affiliation based on *shachôkai* membership and the percentage of client firm shares held by banks

	KEIRETSU	
	0	**1**
Number of observations	237	131
Mean of SHARE	22.55	21.14

Source: KAISHA database, authors' calculation; Oneway-Anova-Analysis: p=0.0135; n=368

Table 3-17: Interrelation between the index of bank-dispatched directors and the percentage of bank debt

	DISPATCH		
	0	**< 10**	**≥ 10**
Number of observations	138	172	58
Mean of bank debt to total debt in %	38.42	47.95	48.51

Source: KAISHA database, authors' calculation; n=368

Table 3-18: Interrelation between the percentage of client firm shares held by banks and the percentage of debt provided by banks

	SHARE		
	< 10	**10 ≥ x < 25**	**≥ 25**
Number of observations	17	222	129
Mean of bank debt to total debt in %	57.82	50.53	31.81

Source: KAISHA database, authors' calculation; n=368

Table 3-17 shows the interrelation between the degree of influence that bank-dispatched directors have on either of the two boards on the one hand, and the percentage of debt provided by banks on the other. Agency theory points out that asymmetries of information between the creditor and the debtor can lead to credit rationing. Considering the aforementioned access to information via bank-dispatched directors, sending one or more representatives to the board of directors and/or the *kansayaku* board is a straightforward way to circumvent this problem. Indeed, we find that the dispatch of one or more bank representatives certainly goes along with a higher percentage of debt provided by banks. The Oneway-Anova-Analysis comes up with a high level of significance (p=0.0000) for the interrelation between DISPATCH and the percentage of debt provided by banks, thus further supporting our finding. Provided the leverage and access to information for banks due to their shareholdings, an increased amount of shareholdings should circumvent information asymmetries similar to the dispatch of bank representatives.

However, table 3-18 suggests that in comparison to firms where banks hold a substantial portion of shares, firms receive a much higher percentage of their loans from banks when banks hold only a few or even no shares. This result can be interpreted in many ways. It is possible that in the case when banks hold a high percentage of shares, they take the shareholder position rather than the debt holder position. Another possibility, which is inherent with all dipolar analyses of this chapter, is that our result is misleading due to interference of other factors that have not been controlled for at the current stage of our investigation.

Summing up, the analysis of interrelations between the different forms of bank-firm relationships used in our study demonstrates that they are not necessarily interdependent of each other. For example, we clearly showed a positive relation between the index reflecting personnel relationships between banks and client firms at board level on the one hand, and main bank affiliation on the other, while the index of personnel relationships does not show any sign of interdependence with the percentage of bank-held shares. The fact that positive and negative as well as no interrelation between the different ways of bank-firm relationship seem to be possible underscores the importance of including all of these types of bank-firm relationships for our analysis. Furthermore, at the current state, the results still leave much room for a wide range of possible interpretations and even speculation. Continuing our study with multivariate analysis based on our panel data set will help us to come up with much more robust findings that shed more light on the advantages and disadvantages of Japanese bank-firm relationships.

Chapter 4

Banks as Monitors

4.1 Introduction

The description of recent changes of Japan's financial markets in general and the role of bank finance in particular (see chapter 2) lead to the generally accepted view that Japanese bank-firm relationships have undergone fundamental changes at least since the early 1980's (BOJ 1992; Kobayashi et al. 1993; Beason and James 1999, among others). However, to conclude that the decline of bank lending gives evidence of a decreasing importance of bank-firm relationships may be too simple. There is no priory reason why a close bank–affiliation should be associated with high bank credits, even though the literature has sometimes implicitly assumed this by using bank debt as a proxy for such relationships.

Indeed, recent economic theory has pointed out that a bank is not only a provider of finance. Financial intermediaries also have to make sure that firms do not squander their capital and therefore have a corporate governance function. In the following (4.2), we will describe this dual role of banks with a reasonably simple model. Similar to Holmström and Tirole (1997), banks can act as monitors of firms in order to control agency problems associated with debt, whereas the investors in the financial markets are too dispersed to invest in monitoring. Firms that take on too much debt in relation to their market value will not have a sufficient stake in the financial outcome and might therefore behave irresponsibly. However, if these firms use a high fraction of bank debt, the bank will monitor them closely. The financial markets will take this into account, allowing the firm access to cheaper market debt. The model shows that the relevant factor for the severity of the moral hazard problem is the ratio of market value to debt and not the more commonly used capital structure. This explains why bank finance was less necessary during the bubble period.

We incorporate bank-firm relationships into our model by assuming that they decrease monitoring costs for the bank. In order to be monitored, a firm with a close bank relationship does not need to take on as much debt as a firm with no such relationship. This seemingly counterintuitive relationship between the fraction of outside bond finance and different proxies for bank-firm relationships can be observed empirically. In this respect our work is part of a growing literature showing that not only the form of the financial contract is important, but also the identity of the lender.

However, we also want to contribute to the literature analyzing the effect of bank-firm relationships in Japan. We cast some doubt on the generally held view that strong bank-firm relationships result in high fractions of bank financed debt. The fraction of bank-financed debt is indeed positively correlated with proxies for close bank relationships, like the intensity of personnel relationships or the amount of bank shareholdings[48]. However, this correlation becomes negative if one controls for firms' total indebtedness. Everything else constant, better bank-firm relationships seem to facilitate firm access to the bond market.

Our interpretation is that in general, bank-firm relationships are a consequence rather than a cause of high indebtedness. Firms with a high need for debt seem especially likely to develop these relationships in order to sustain the level of indebtedness. However, this is not the only reason for close relationships. For firms in the period analyzed, many relationships can be explained exogenously by historical reasons. Being members of a *kinyû keiretsu*, a vast number of bank relationships date back to the 1950s (Yafeh 1995), and many bank affiliated firms even have forerunners that belonged to the pre- World War II *zaibatsu*[49] form of organization (Hoshi 1993). In addition, firms with no *kinyû keiretsu* affiliation may have established a bank relationship due to their need for finance in the past. As these relationships are very stable once established, they alter the way the firms react to more short-term changes in financing conditions, even if the original motivation for the relationship has disappeared. This exogenous variation in bank-firm relationships can be used to empirically separate the effects of such relationships from the effects of high indebtedness compared to the value of the firm. As it turns out, controlling for debt load, bank-firm relationships in Japan facilitate access to the bond markets.

Our results are similar to the observations of Hoshi, Kashyap, and Scharfstein (1991). They show that firms with a *kinyû keiretsu* affiliation tend to have a better capital structure than firms without a close bank relationship. The findings shown below provide evidence that bank relationships seem not only to facilitate access to the bond markets but also access to equity.

We provide evidence that despite the decline in bank lending, Japanese banks still have an outstanding monitoring function in the Japanese financial system and are still important institutions that mitigate problems of asymmetric information. We believe that the predominance of the bank role results from several factors which prevent, despite recent liberalization, the Japanese capital markets from providing effective corporate governance. Especially widespread cross-shareholdings dilute the influence of individual shareholders and prevent the threat of hostile takeovers from being an instrument of corporate control and discipline. With the maturing of the Japanese capital markets, the position of banks will certainly change. Nevertheless, in the intermediate phase between a bank-

---

48 See chapter 3 for a further assessment of bank-firm relationships

49 A *zaibatsu* is a group of diversified businesses owned exclusively by a single family or extended family (Morikawa 1992 p. xvii).

dominated and a market-oriented financial system, banks still seem to serve an important function as a provider of capital as well as of corporate governance.

The rest of the chapter is organized as follows. In 4.2 we introduce a model as the basis of our analysis to explain the role and impact of bank-firm relationships as well as the firm's market value in the changing patterns of corporate financing. The empirical analysis that follows in 4.3 provides different measures to analyze bank-firm relationships. Furthermore, the testable implications of the model will be examined using data from 92 large Japanese manufacturing firms for the period between 1985–1998 (see chapter 3). The chapter concludes with some final remarks in 4.4.

4.2 The Model

In the following we build a reasonably simple microeconomic model of outside finance to describe the role of banks as providers of finance and corporate governance. The model accounts for much reported evidence that firms with a high market value can choose less information-intensive, (direct) outside financial sources (e.g. BOJ 1992, Shimizu 1997a). These sources are generally perceived to be cheaper than (indirect) bank finance[50]. Even though bank debt is generally more expensive than other sources of outside finance (bonds), the firm demands a fraction of bank-financed debt in order to be monitored by the bank. The more the bank is to monitor the firm, the more debt the firm has to demand from the bank to give the bank an adequate incentive to increase the intensity of monitoring. The larger the incentives for the firm's entrepreneurs to commit moral hazard, the stronger the intensity of monitoring has to be. This again implies that the firm has to demand a high enough fraction of bank-debt to ensure that the bank still monitors the firm sufficiently.

We assume that close bank-firm relationships reduce the marginal costs of monitoring. This implies that lower incentives for the bank are sufficient to achieve the same level of monitoring. In other words, with close relationships, lower bank debt leads to the same monitoring intensity, i.e. the smaller the marginal costs, the smaller the need for bank-financed debt.

50 "...the nominal issuing cost of convertible bonds and bonds with warrants declined sharply in the second half of the 1980s, as the rise in stock prices and the resultant expectations of a continues growth in stock prices raised the value of options included in these bonds; and corporate managers ... seemed to have perceived the issuing cost to be only the coupon rates of these bonds and thus judged the level of cost to be extremely low" (Bank of Japan 1996, pp. 64-65).

4.2.1 The Setting of the Game

In the model we consider three types of agents: (1) firms, represented by the index F; (2) banks which have the ability to monitor, represented by the index B; and (3) competitive investors organized in a financial market, represented by the index M. All agents are risk-neutral. To keep the model reasonably simple, there are only two sources of outside finance: one is bank debt, representing indirect, intermediated and monitored finance, and the other is bonds, representing direct market finance from competitive investors. The banks have access to an alternative investment opportunity yielding an expected rate of return of r_B, corresponding to an interest rate of $(r_B - 1)$. The investors' opportunity costs are r_M, with $r_B > r_M$.

Each firm has an investment project that produces a return of R in case of success and nothing if the project fails. The amount of capital a firm needs for its investment is normalized to 1. In order to capture the moral hazard problem of the firms in a simple way, we assume that the firms have the choice between a safe and a risky business strategy. The safe strategy has a low risk of default, implying a high probability p_h of success; the risky strategy leads to a lower success probability of p_l, i.e. $\Delta p = p_h - p_l > 0$.

The safe strategy is socially optimal; however, the high-risk strategy gives the entrepreneur a private benefit $b(m)$ in addition to his financial remuneration. If this private benefit exceeds his financial losses from choosing the risky project, he may intentionally choose the inefficiently low probability p_l of success. We assume that the private benefit $b(m)$ can be decreased by bank monitoring where m represents monitoring intensity. Thus the bank can influence the strategic choice of the firms by making it more difficult to enjoy private benefits.

There are several ways to interpret the private benefits in our model in the light of more elaborate principal-agent theories of financing. If we assume that the firm is run by a manager-owner, the private benefits can be understood as the classical asset-substitution moral hazard effect (Jensen and Meckling 1976; Stiglitz and Weiss 1981). As the firm does not pay for the negative aspects of incurring risk, increasing the variance of the output without decreasing the expected output too much will increase the owner's expected profit[51]. A monitoring bank can render this kind of risk-increasing moral hazard unattractive by threatening to withdraw credit, reacting with an increase in interest rates (Tröge 1999) or directly preventing risky strategies through its influence channels. Moreover, the private benefits in our model can also be understood as a negative effort in a principal-agent setting. This is especially relevant if management controls the firm without having strong incentives to defend primarily shareholder interests. A monitoring bank is able to detect if the firm's manager is not working efficiently and take precautions such as firing the manager, implementing lower

[51] Note that in our setting, increasing the variance i.e. decreasing p will not increase the owner's profit as the expected output will decrease at the same time.

salaries etc.

However, monitoring is also costly for the bank. Having only a small credit engagement, a bank will probably not find it worthwhile to invest in information acquisition, but rather to compensate the risk by asking for a higher interest rate. We assume that a close relationship between the bank and the firm will reduce the cost of monitoring for the bank.

The financial markets will only be willing to provide cheap finance if the firm chooses the strategy with a high probability p_h of success. However, it is assumed that neither the strategic choice of the entrepreneur nor the monitoring activity of the bank can be contractually enforced. Therefore there are only two conditions under which the financial markets will be convinced that the firm is not going to follow a risky strategy: Either the firm's prospects are sufficiently good (indicated by a high market value) and its capital is sufficiently high so that it would not be worthwhile for the firm to increase risk, or investors know that the firm is monitored by a bank and therefore private benefits are too low to make the risky strategy profitable.

Investors will not believe that the bank is monitoring appropriately, however, if the bank has no incentives to monitor. Monitoring is profitable for the bank only if the firm has a sufficiently high share of bank financing. Therefore, in order to incite the bank to monitor and to convince outside investors that it does not pursue a risky strategy, a firm with a high need of finance compared to its market value has to ask the bank for a sufficiently high share of credits.

We model the decision process as a three-step optimizing problem. First, the firm can decide about the fraction of bank-financed debt λ_B. The market will finance the remaining

$1-\lambda_B$. Depending on the expected risk and the required rate of return, the bank and the financial market ask for a repayment of R_B and R_M respectively in case of success. It is presumed that R_B, R_M are competitively determined. The remaining return for the firm is therefore $R_F = R - R_B - R_M$. We assume that $Rp_l < 1$, i.e. the project with low probability of success has a negative net present value.

For the monitoring function $b(m)$, we choose a linear specification, i.e. $b(m) = b(0) - am$. It is assumed that investing one unit in monitoring reduces the private benefit by more than one unit, which means that $a > 1$, and that the private benefits are sufficiently large, so that without monitoring the risky project is always chosen. For simplicity, the cost of monitoring is assumed to be linear, i.e. $c(m) = cm$. The bank's cost c to achieve a monitoring intensity of m depends on the intensity of the relationship with the firm. We assume that $c \in [0,1]$. A bank without a relationship has to pay the marginal cost of $c = 1$ in order to monitor; closer relationships go along with smaller c.

To summarize, the sequence of the game is as follows:

1. The firm determines λ_B
2. The bank determines the monitoring intensity m
3. The firm chooses the success probabilities $p \in \{p_h, p_l\}$

4.2.2 Equilibrium strategies

We solve the game by backward induction.

The firm's problem at stage three In the last stage R_M, R_B and m have already been determined in earlier stages of the game. The firm's profit is:

$$(1)\ \max\{R_F p_l + b(m), R_F p_h\}$$

Depending on which strategy choice maximizes profit, the firm decides to take the risky or the safe business strategy. If choosing the high-risk strategy, the firm will receive lower expected returns but also the private benefit of $b(m)$. It is obvious that the firm has to decide on a trade-off between $b(m)$ and the increase $R_F \Delta p$ of residual payments if it chooses p_h. With a high monitoring intensity, private benefit is low; therefore the firm chooses the low-risk project. If the bank does not monitor sufficiently, the firm will choose the business strategy with a low probability of success but a high degree of private benefits. We assume that the firm chooses the efficient solution if it is indifferent. Let $m^*(R_F)$ be the level of monitoring intensity that makes the firm indifferent if in case of success it receives a residual payment R_F. This is the smallest monitoring intensity such that $R_F \Delta p = b(m^*)$. Since $b(m) = b_0 - am$, we get $R_F \Delta p = b_0 - am^*$, so that

$$(2)\ m^* = \frac{b_0 - R_F \Delta p}{a}$$

We define the success probability as a function of the monitoring intensity:

$$(3)\ p(m) = p_h \text{ for } m \ge m^*$$
$$= p_l \text{ for } m < m^*$$

We have $\frac{\partial}{\partial R_F} m^* = -\frac{\Delta p}{a} < 0$.

This means the larger the return going to the firm, the less monitoring is needed to prevent it from choosing the risky strategy.

The bank's problem at stage two In the second stage, the bank knows the return R_B it receives in case of success. The success probability, however, depends on its monitoring effort. The bank therefore maximizes:

$$(4)\ \max_m \{R_B p(m) - c(m)\}$$

The incentive mechanism for the bank is similar to that of the firm: Acting is costly, so the bank only acts if it gets a sufficiently large share of the return in case of success. If c is low, it will be worthwhile for the bank to reduce the private benefits of the entrepreneur sufficiently to implement the choice of the safe project. In this case the optimal monitoring level is the lowest monitoring level $m^*(R_F)$ for which this choice will be implemented. If implementing the safe project is too expensive, it will be optimal not to monitor at all. Hence if $R_B \Delta p \geq c(m^*)$, the bank chooses the monitoring level m^*, otherwise it does not monitor. If the bank is indifferent between $m = 0$ and m^*, we assume that the bank chooses m^*.

The firm's choice of its debt structure at stage one The firm gives the bank the incentive to monitor in order to commit itself not to take the risky action by choosing the appropriate capital structure. The firm can choose whether the project will be entirely bond financed or if a fraction will be bank-financed. The latter capital structure is only profitable if the banks are given an adequate incentive to monitor sufficiently. As we assume that the low success project is not profitable, in this case the project will not be financed. We know that the bank will only monitor if, in case of success, it gets a return greater than or equal to:

$$(5)\ R_B = \frac{cm^*}{\Delta p}$$

If the banks are ex ante competitive and need a return of r_B, they will be willing to provide an amount of capital up to λ_B for a payment of R_B in case of success, which can be calculated as follows:

$$(6)\ r_B\left[\lambda_B + cm^*\right] = p_h R_B$$

$$\Rightarrow \lambda_B = \frac{p_h}{r_B} R_B - cm^* = \left(\frac{p_h}{r_B \Delta p} - 1\right) cm^*$$

It would be excessively costly to have more bank finance, as the required return on bonds r_M is smaller.

If $\left(\frac{p_h}{r_B \Delta p} - 1\right) cm^* > 1$, the project will be entirely bank-financed. If $\frac{p_h}{r_B \Delta p} < 1$, bank finance is too expensive and the project will not be financed at all.

Since all firms in our sample have been financed we can assume that $\frac{p_h}{r_B \Delta p} \geq 1$.

The success return that has to be paid to the bondholders is

$$(7)\ R_M = \frac{r_M}{p_h}\left(1 - \lambda_B\right) = \frac{r_M}{p_h}\left[1 - \left(\frac{p_h}{r_B \Delta p} - 1\right) cm^*\right]$$

$$= r_M\left[\frac{1 + cm^*}{p_h} - \frac{cm^*}{r_B \Delta p}\right]$$

If the investors observe a lower fraction of bank-financed debt, they assume that the firm will not be monitored. This will imply the choice of the unprofitable project; therefore the market will not choose to provide finance.

4.2.3 General Properties

After the solution has been traced by backward induction, it is possible to derive the general properties of the model.

Differentiating equation (5) with respect to R_F yields:

$$(8)\ \frac{\partial}{\partial R_F} R_B = \frac{c}{\Delta p}\frac{\partial}{\partial R_F} m^* = -\frac{c}{a} > -1$$

If the residual return to the firm increases by one yen, the firm can reduce the bank's share of the success payoff by less than one yen.

From equation (7) we get:

$$(9)\ \frac{\partial}{\partial R_F} R_M = r_M\left[\frac{1}{p_h} - \frac{1}{r_B \Delta p}\right] c \frac{\partial}{\partial R_F} m^* > 0$$

We have taken the assumption made in 4.2.2, that bank finance is not too expensive. The market's share of the success payoff will increase with that of the firm.

From equations (8) and (9) we can derive:

$$(10)\ \frac{\partial}{\partial R_F}(R_B + R_M) = \left[\frac{1}{\Delta p} + \frac{r_M}{p_h} - \frac{r_M}{r_B \Delta p}\right] c \frac{\partial}{\partial R_F} m^* > -1$$

As the firm's share of profit increases, monitoring becomes less necessary. Hence the return that has to be paid to the bank decreases faster than the return for the bondholders increases. It is straightforward that an increase in the gross value of the firm will lead to an increase in the firm's share of the profit. Formally, this can be shown by differentiating the equation:

$$(11)\ R = R_F + R_M(R_F) + R_B(R_F)$$

with respect to R :

$$(12)\ 1 = \frac{\partial R_F}{\partial R} + \frac{\partial R_M(R_F)}{\partial R_F}\frac{\partial R_F}{\partial R} + \frac{\partial R_B(R_F)}{\partial R_F}\frac{\partial R_F}{\partial R}$$

$$\Rightarrow \frac{\partial R_F}{\partial R} = \frac{1}{1 + \frac{\partial (R_M + R_B)}{\partial R_F}} > 0$$

An increasing R corresponds to an increase of R_F .

4.2.4 Testable Implications

We are interested in two relationships: How does an increase or decrease of $\boldsymbol{R}$ and c affect λ_B? We have already shown that

$$(13)\ \lambda_B = \frac{p_h}{r_B} R_B - cm^* = c\left(\frac{p_h}{r_B \Delta p} - 1\right) m^*$$

Differentiating with respect to $\boldsymbol{R}$ yields:

$$(14)\ \frac{\partial}{\partial R}\lambda_B = \frac{\partial}{\partial R_F}\lambda_B \frac{\partial}{\partial R} R_F = c\left(\frac{p_h}{r_B \Delta p} - 1\right)\frac{\partial}{\partial R_F} m^* \frac{\partial}{\partial R} R_F < 0$$

Differentiating with respect to c yields:

$$(15)\ \frac{\partial}{\partial c}\lambda_B = \left(\frac{p_h}{r_B \Delta p} - 1\right)\frac{\partial}{\partial c} cm^* = \left(\frac{p_h}{r_B \Delta p} - 1\right) m^* > 0$$

again assuming $\lambda_B > 0$.

The intuitive meaning of these results should be clear. An increase in the firm's market value as well as a close relationship with a bank decreases the entrepreneurs' incentives for risky behavior and enables them to take on more external finance. The former holds, since with a high market value the owner-entrepreneur bears a greater share of the costs of risky behavior. For the latter result we have assumed that close relationships lead to increased monitoring, which decreases the private benefits of risky behavior.

We can reformulate these results in two hypotheses:

1. A higher value of the firm leads to a lower fraction of bank finance.
2. Close bank-firm relationships lead to a lower fraction of bank finance.

Note that alternative explanations could be imagined. However, in chapter 5 we further test the direct impact of bank-firm relationships on firms' input- as well as output-factors. The results obtained further reinforce the view explained above.

The model only allows us to make these qualitative predictions without providing precise functional forms. Therefore we will test the qualitative implications in a linear specification, which we will discuss in 4.3. As the market value of firms is strongly negatively correlated with variables that had been used in earlier works to indicate close bank-firm relationships (main bank dummies, etc.), it is necessary to test both hypotheses in a single equation. Earlier papers running regressions between proxies of bank relationships and the fraction of bank-financed debt have neglected to correct for the effect of market value and therefore obtained a positive relationship. Running the regression with both explanatory variables, it becomes clear that this results form the indirect effect of bank relationships on market value and of market value on the fraction of bank finance. As this effect is much stronger than the direct effect, it dominates in the regression. Including the market value in the regression enables us to separate the two factors and to isolate the effect of bank relationships. It will be shown in the following that their influence is indeed negative. Close bank relationships facilitate the access to the bond market.

4.3 Empirical Analysis

4.3.1 Data Description and Estimation Procedure

For our empirical analysis, we use the data set of 92 sample firms over the years from 1985-1998, which has already been described in chapter 3. Following the model, the costs of monitoring certainly depend on the degree and nature of established bank-firm relationships. The closer an established relationship, the smaller are the marginal costs for a bank to monitor its client-firm. As has been already pointed out in chapter 3, there are numerous ways to assess the different degrees and nature of these relationships. For the empirical analysis discussed in this chapter, out of the four measures introduced in chapter 3, we used the index expressing personnel relationships between banks and client firms (DISPATCH), as well as the measure representing the percentage of client-firm-shares held by banks (SHARE)[52]. These measures include different aspects and degrees of bank-firm relationships and provide evidence for marginal costs of monitoring which have been mentioned in the model.

4.3.2 The Empirical Model

4.3.2.1 Benefits of Panel Data Analysis

The data description laid out in chapter 3 explained that our data set follows a given sample of 92 firms over an entire sample period from 1985-1998. To test for poolability of the data used in the following analysis, as well as in analyses discussed in chapters thereafter, we conduct a "Chow test", better called a test of

---

52 We had also included the measure that captured the degree of a particular *kinyû keiretsu* affiliation under the aspect of loan-dependency (KEILO), as well as dummy variables for either *keiretsu* or main bank affiliation in initial modeling attempts (KEIRETSU and MB). However, these variables were not significant over the total period observed. Hence, we do not include them here.

parameter invariance (see Chow 1960; Baltagi 1995). We succeed in accepting the hypothesis of parameter invariance, thus allowing us to pool the data. Furthermore, there is no evidence for parameter heterogeneity. In other words, our data set provides multiple observations on each firm in the sample. Summing up, we can conclude that the data set meets all necessary conditions to be a true balanced panel data set.

To have a firm-level panel data set at hand grants us several major advantages over previous empirical analyses concerning Japanese bank-firm relationships. Earlier analyses only used conventional cross-sectional data sets, time-series data sets, or did not use the full range of advantages a panel data analysis offers (Nakatani (1984); Hoshi, Kashyap and Scharfstein (1991); Weinstein and Yafeh (1998), among others).

Highlighting only the most obvious advantages for our study to use panel data analysis, following Hsiao (1986) and Baltagi (1995), the two primary advantages are the following: First, our panel data set dramatically improves the efficiency of econometric estimates, since it provides us with a large number of data points, increases the degrees of freedom and reduces the collinearity among explanatory variables. Therefore, it helps us to determine the individual influence of each explanatory variable. Second, it enables us to expand our study by analyzing aspects that cannot be addressed by using conventional cross-sectional or time-series data sets.

To give an example closely related to our study, assume that we conduct an analysis concerning bankers at the board of directors by using a cross-sectional sample. Suppose the analysis of the boards finds that of all sample firms, an average yearly rate of fifty percent have a banker on the board of directors. Though this result might appear to be clear cut at first sight, it leaves room for a wide range of interpretations, which can cause great difficulties of making inference about the dynamics of change from cross-sectional evidence. Underlining the variety of conceivable interpretations, let us take two possible extremes: On the one hand, the result can be interpreted as implying that each firm in a homogeneous community of firms has a fifty per cent chance of having bank-dispatched directors on the board in any given year. In this case, each firm would be expected to have a banker on the board during half of all years the firm exists, and half of the years it has no banker on the board. The turnover of having a banker on the board would be expected to be frequent, with an average duration of two years for having a banker on the board. However, on the other hand the findings might imply that fifty per cent of the firms' board of directors in a heterogeneous community of firms always have bank-dispatched directors on the board, while the other fifty per cent never have bank representatives on the board. In this case, there is no turnover, and current information about the board status (with or without a banker) is a perfect predictor of future work status (see Ben-Porath (1973) for an example on labor force participation rates of married woman that addresses a similar interpretation problem).

Considering the previous example, it is obvious that in order to ensure the validity of our findings, we have to be able to discriminate between the two described models. This is possible with panel data analysis. Having sequential observations for a number of individual boards of firms at hand, it is possible to use these individual board histories concerning the participation of bank-dispatched members to estimate the probability of participation of bankers on boards in different subintervals of a firm's life cycle.

Equally important, due to the lack of firm-level panel data, it was not possible for a large number of previous studies to construct a proper recursive structure in order to observe before/after effects (e.g. change from a bank-affiliated status to a non-affiliated status, or vice versa). We adapt an example by Hsiao (1986) to our specific research interest, in order to underline the consequences of this structural data problem for analyzing aspects of bank-firm relationships. Consider the impact of bank-affiliation on corporate performance. Those economists who tend to interpret the observed differences between affiliated and non-affiliated firms as largely real believe that the potentials of bank-influence fundamentally alter the key aspects of corporate performance: the approach towards risk, the use of capital, the probability of failure, etc. Those economists who regard bank-affiliation effects of some sort as largely illusory, tend to posit that the real world is close enough to satisfying the conditions of perfect competition; they believe that the observed affiliation/non-affiliation differences are mainly due to differences between affiliated/non-affiliated firms prior to bank-affiliation or post-affiliation sorting. If one subscribes to the former view, the coefficient of the dummy variable for bank-affiliation in a corporate performance equation is a measure of the effect of bank-affiliation. If one subscribes to the latter view, then the dummy variable for bank-affiliation could be simply acting as a proxy for firm quality. A single cross-sectional data set usually cannot provide a direct choice between these two hypotheses, because the estimates are likely to reflect inter-firm differences inherent in comparison of different firms rather than intra-firm dynamics, unless data on variables controlling for these differences are available and are explicitly included in the chosen specification.

However, with our panel data set it is possible to follow our 92 firms over time as they change their status. In contrast to Nakatani's (1984) study about corporate grouping, for example, where firms are only included which have been continuously affiliated with a particular group, our data set allows us to distinguish inter-firm from intra-firm differences and to construct the proper recursive structure needed for our study of a before-and-after effect. For example, using our sample we can examine the development of performance differentials of firms moving from a bank-affiliated status to a non-affiliated status (and vice versa). Thus, even if specific information on firm quality is not available, if the firm's quality stays constant or changes only slowly, the within correlation between the affiliation-status dummy and the firm-quality variable is likely to be negligible. In other words, firm quality can be controlled through the use of within estimates. The resulting coefficient for the affiliation-status dummy (e.g. main bank-affiliated or non-affiliated) will then provide a measure of the impact of main bank-affiliation. Recalling the dramatic changes of bank-firm relationships during the period under observation, there is no doubt about the importance of being able to compare the *same* set of firms over a given period rather than running the risk that estimates are likely to reflect inter-individual differences inherent in comparisons of *different* firms.

Due to problems of collinearity, among a number of other problems, time-series data usually cannot provide precise estimates of dynamic coefficients either and are not particularly useful for discriminating between hypotheses that depend on microeconomic attributes (see Hsiao (1986) for more detailed discussion).

Finally, since it is possible to utilize information on the inter-temporal dynamics as well as the individuality of the firms under investigation, using panel data analysis enables us to control much better for effects of missing or unobservable variables. Hence it allows us to mitigate the econometric problem that the real reason one finds certain effects is because of omitted variables that are correlated with explanatory variables. If the effects of omitted variables are correlated with the explanatory variables included in an estimation and if these correlations are not explicitly allowed for, the resulting estimates will be biased (see Hsiao 1986). Using panel data analysis, we can minimize the bias since it is possible to distinguish different types of correlations between the included variables and the error term. Distinguishing the different sources of correlations allows us to gain important information for constructing consistent estimators. For instance, in the estimations discussed below we were able to use the dummy-variable (fixed-effects) approach to capture the effects of firm-invariant and time-invariant variables, therefore eliminating the omitted-variable bias. In other words, compared to other empirical analyses concerning Japanese bank-firm relationships, using panel data analysis we have a great advantage in facing the well known specification problem.

4.3.2.2 Methodology of the Empirical Analysis

It is most obvious that using a balanced panel data set provides major benefits to our empirical study. However, out of the various econometric methods available for panel data analysis, we have to choose the most appropriate one for exploiting the richness and unique properties of our data set. In addition, we have to be aware of the numerous limitations of the features of panel data in general and of each specific method in particular.

Since the test of joined significance of time dummies showed the significance of time-effects, we decided to account not only for individual firm effects, but also for any individual-invariant, time-specific effect that is not included in our regression. Hence, following Wallace and Hussain (1969), Nerlove (1971), and Amemiya (1971), we use a regression model with two-way error components disturbances of the type:

$$(16)\ y_{it} = \alpha + X_{it}^{'}\beta + u_{it} \qquad i = 1,...,N;\ t = 1,...T$$

with

$$(17)\ u_{it} = \mu_i + \lambda_t + v_{it}$$

In our case, i denotes firms and t denotes time. α is a scalar, β is $K \times 1$ and X_{it} is the itth observation on K explanatory variables. μ_i denotes the unobservable individual firm effect, λ_t denotes the unobservable time effect and v_{it} is the remainder stochastic disturbance term representing the effects of those unobserved variables that vary over i and t (see Hsiao 1986, Baltagi 1995).

Whatever the properties of μ_i, λ_t and v_{it}, if equation (16) is true, it must also be true that

$$(18)\ \overline{y_{i.}} = \alpha + \overline{X_{i.}}\beta + \mu_i + \overline{v_{i.}}$$

where $\overline{y_{i.}} = \sum_t y_{it} / T_i, \overline{X_{i.}} = \sum_t X_{it} / T_i$, and $\overline{v_{i.}} = \sum_t v_{it} / T_i$.

Furthermore,

$$(19)\ \overline{y_{.t}} = \alpha + \overline{X_{.t}}\beta + \lambda_t + \overline{v_{.t}}$$

where $\overline{y_{.t}} = \sum_i y_{it} / N_t, \overline{X_{.t}} = \sum_i X_{it} / N_t$, and $\overline{v_{.t}} = \sum_i v_{it} / N_t$.

One can now deduce that

(20)

$$\left(y_{it} - \overline{y_{i.}} - \overline{y_{.t}} + \overline{y_{..}}\right) = \left(X_{it} - '\overline{X_{i.}} - \overline{X_{.t}} + \overline{X_{..}}\right)\beta + \left(v_{it} - \overline{v_{i.}} - \overline{v_{.t}} + \overline{v_{..}}\right)$$

Ordinary least squares on this model gives $\tilde{\beta}$, the Within estimator for the two-sided model.

When inferences are going to be confined to the specific units in the model, the effects are more appropriately considered fixed. Using a fixed-effects model, we make inferences conditional on the effects that are in the sample. However, there is loss of efficiency in conditional inference because of the loss of degrees of freedom in estimating the specific effects. When inferences will be made about the population, the effects are more appropriately considered random. Using a random-effects model, we make unconditional or marginal inference with respect to the population of all effects. Nevertheless, one has to keep in mind that a random-effects inference requires a distributional assumption with regard to the effects. For instance, when using a linear model, the assumption can be in the general form of independently distributed effects, with common mean and finite variance-covariance matrix (see Hsiao 1986; Baltagi 1995 or Greene 1993, among others for more detailed discussion).

Analyzing Japanese bank-firm relationships, generally speaking, it is up to the investigator to decide whether to make inference with respect to the general characteristics of all Japanese firms or only with respect to the effects that are in the sample. However, when deciding whether a random-effects or a fixed-effects model is more appropriate, we should take the context of the data as well as a test for misspecification into account. Considering our sample design (see chapter 3) and the aspects under observation, the inferences should be confined to the effects of the model; thus a fixed-effects model appears to be more appropriate. Having decided that μ_i and λ_t are assumed to be fixed parameters to be estimated and the remainder disturbances stochastic with $\nu_{it} \sim \boldsymbol{IID}(0, \sigma_\nu^2)$, we can address the regression model (16) a two-way (or two-sided) fixed effects error component model, where $\boldsymbol{X}_{it}$ are assumed independent of the ν_{it} for all $\boldsymbol{i}$ and $\boldsymbol{t}$. Note that in this case, as has already been pointed out above, inference is conditional on our 92 firms and over the period between 1985-1998.

To test for misspecification and therefore further examine the appropriateness of a fixed rather than a random-effects model, we followed Kang (1985) based on Hausmann's (1978) specification test to consider the implications of our two-way model (for an alternative testing procedure, see Hsiao 1986:48). Identifying a strong indication of misspecification in the conventional random-effects model[53], we can conclude that important firm-specific as well as time-specific effects are present that are correlated with the right-hand variables (a discussion on the specific variables used will be provided below). Thus, as has been pointed out in a similar case by Hsiao (1986), because the random-effects estimates appear to be significantly biased with high probability, it seems to be important to take permanent unobserved differences across firms and years into account when estimating our equations by using panel data. For our following linear-regression models (in chapter 4 and 5) we therefore restrict our analysis to marginal inferences that properly allow for the correlation between individual and time effects and the attributes generating the process. Furthermore, the analyses need not make inference about population characteristics, thus assuming a fixed-effects model.

4.3.3 Empirical Results

To test the model introduced in 4.2, we observe in our empirical analysis the implications of the firm's market value and the degree and nature of bank-firm relationships on the firm's decision regarding the fraction of bank-financed debt. In the model we assume that the firm invests one unit of capital in the form of debt. For our empirical analysis, we therefore construct the variable that accounts for the firm's market value compared to the firm's total liabilities (MV). The measures DISPATCH and SHARE, which have already been introduced in chapter 3,

---

[53] The $\boldsymbol{F}$ value computed by Haussman (1978) is 137.8 , which well exceeds the one per cent critical value.

account for the different degrees and nature of bank-firm relationships[54] and therefore give evidence for marginal costs of monitoring. Using factor analysis, we furthermore developed a measure comprising the accessibility and actual usage of bond-financing (ACCESS). This measure has been introduced since there are a variety of additional sources of outside finance besides bank- and bond-financing available, not included in the model. We use the firm's credit rating to account for bond accessibility[55]. The number of newly issued bonds[56] is used to assess the actual usage of bond-financing. The dependent variable, which is the fraction of bank-financed debt, is simply the amount of bank debt normalized by total liabilities (for a definition of variables and descriptive statistics, see appendix I and II).

Observing the parameter for the market value of the firm (MV), the results represented in table 4-1 show that the fraction of bank-debt is determined by the firm's market value. As predicted by the model introduced in 4.2, our empirical evidence suggests that the higher the market value of the firm per unit of debt, the smaller is the fraction of bank-financed debt. We can therefore claim the first hypothesis to be true. As expected, the result of the parameter ACCESS underlines that the availability and usage of bond-financing clearly has a negative effect on the firm's demand for bank-financed debt.

Table 4-1: Regression on bank monitoring, 1985-1998

Dependent variable: bank debt normalized by total liabilities
Fixed effects regression coefficients (T-values)

		(1)
Relationship Variables	DISPATCH	-0.00133*** (-2.682)
	SHARE	-0.00110** (-2.191)
	Sig. relationship variables	0.0000

Note: *, **, ***: Significance at the 10%, 5% and 1% level, respectively.

54 The correlation coefficient between DISPATCH and SHARE is –0.0564, which does not indicate a multicollinearity problem.

55 We would like to thank Bloomberg L.P. for providing this data.

56 With bonds we mean convertible bonds, bonds with warrants and straight bonds.

Table 4-1: Regression on bank monitoring, 1985-1998 (continued)

	Fixed effects regression coefficients	T-values
Control Variables	MV	-0.02141*** (-15.495)
	ACCESS	-0.06867*** (-42.878)
	Yr1986	0.01725** (2.226)
	Yr1987	0.0313*** (4.064)
	Yr1988	0.03070*** (3.710)
	Yr1989	0.02399*** (2.836)
	Yr1990	0.01492* (1.775)
	Yr1991	0.00317 (0.384)
	Yr1992	0.00868 (1.034)
	Yr1993	0.01868** (2.231)
	Yr1994	0.03710*** (4.422)
	Yr1995	0.02882*** (3.443)
	Yr1996	0.03386*** (4.079)
	Yr1997	0.02345*** (2.833)
	Yr1998	0.02105** (2.545)

Note: *, **, ***: Significance at the 10%, 5% and 1% level, respectively.

Table 4-1: Regression on bank monitoring, 1985-1998 (continued)

Statistics	N of observations	1288
	N of groups	92
	Obs. per group	14
	Sign. F	0.0000
	R^2	0.6492
	DW[57]	1.9534

The parameters of the two different measures of bank-firm relationships (DISPATCH and SHARE) are significant at a 5 per cent or even 1 per cent level. The results show that the closer the bank-firm relationship, the smaller is the fraction of bank-financed debt; hence, we were also able to prove the second hypothesis to be true. The results thus clearly support the predictions of our model.

4.4 Conclusion

With a simple model of bank monitoring we have derived the counterintuitive but empirically observable finding that, everything else constant, close bank-firm relationships will lead to a lower fraction of bank finance. This provides evidence that Japanese banks take an important role in the corporate governance of large firms.

The predominance of banks as providers of corporate governance in Japan may be the result of several factors which prevent, despite recent liberalization, the evolution of other monitoring institutions. Especially widespread cross-shareholdings dilute the influence of individual shareholders and prevent the threat of hostile takeovers from being an instrument of corporate control and discipline. With the maturing of the Japanese capital markets, the position of banks may change.

However, it may also be that banks have a natural and privileged position for providing efficient corporate governance. Strict regulation and liability requirements make it still difficult for Japanese banks to carry out this function. Further liberalization of Japan's banking regulation may change this.

---

57 For Durbin-Watson, here as well as for the analyses below, we followed Bhargava, Franzini, and Narendranathan (1982), who generalize Durbin-Watson type statistics to test the OLS residuals from the fixed effects model for serial independence.

Chapter 5

The Recent Economic Role of Bank-Firm Relationships

5.1 Introduction

The following analysis is motivated by theories of financial intermediation that accentuate the information advantages of financial intermediaries, through which they can (partially) overcome frictions such as information asymmetries and agency problems of managerial behavior (e.g. Leland and Pyle 1977; Campbell and Kracaw 1980; Diamond 1984, 1991; Ramakrishnan and Thakor 1984; Fama 1985; Boyd and Prescott 1986; Haubrich 1989; Rajan 1992). As has already been pointed out in previous chapters, especially in financial systems like those of Japan or Germany, banks may acquire information through their relationships with firms by continuously monitoring the performance of clients under credit arrangements, as well as through the provision of other services such as deposit accounts, cash management services, bankers' acceptances, credit card processing, pension fund management, factoring, or sales financing. Banks may enjoy economies of scale as well as comparative advantages in the production of information about borrowers. Being a debt holder as well as a shareholder that in many cases maintains personnel ties with a client firm, bank influence is not limited to indirect influence from information acquired for credit decisions. The banks' position enables them to influence the corporate behavior of a client firm directly, for example, by using their voting rights as a major shareholder or by implementing decisions via their representative(s) on the board of the firm.

Due to the banks' unique position, contracting in the bank loan market seems to differ substantially from contracting in other debt markets that offer arm's-length finance[58]. In the last decade, empirical literature drawing from US as

---

[58] As has been already pointed out in an early version of Berglöf and Sjögren (1998), the choice of typology is somewhat arbitrary. Holmström (1992) and Holmström and Tirole (1993) use the terms "intermediated" and "informed" interchangeably, for what is here called intermediary or bank finance, and "uninformed" for arm's-length finance. Berglöf and Sjögren (1998) use the term control-oriented finance rather than intermediary finance. None of these terms are ideal. Monitoring is just one form of financial intermediation. The dichotomy "uniformed-informed" highlights on the information available to investors whereas the crucial distinction is whether or not investors influence investment decisions. Also, "uninformed" connotes that assessing collateral requires no information. However, collateral requires that information is

well as European[59] data has further examined the exceptional features of intermediaries in general, and bank-firm relationships in particular (e.g. James 1987; Lummer and McConnell 1989; James and Wier 1990; Wansley, Elayan, and Collins 1992; Shockley and Thakor 1993; Kwan 1994; Petersen and Rajan 1994, 1995; Berger and Udell 1995, Billett, Flannery and Garfinkel 1995; Elston and Albach 1995; Albach 1998, 1999b; Harhoff and Körting 1998a, 1998b; Berglöf and Sjögren 1998; Degryse and Ongena 1999; Yang 2000). Following the same strain of literature but examining Japanese competitiveness, Nakatani (1984); Hoshi, Kashyap, and Scharfstein (1990a, 1990b, 1991); Teranishi (1993); Aoki, Patrick, and Sheard (1994); as well as Hoshi, Kashyap, and Loveman (1994), among others, claim that close relationships between firms and financial institutions are important to the international competitiveness of large Japanese firms.

Regardless of whether Japanese, U.S. or European data is examined, most of these studies indicate, among other things, the advantages of relationships between banks and firms, especially when it comes to borrowing. In particular, the majority of the analyses agree that the primary benefit to a firm that builds close ties with one or more banks is an increase in the availability of financing[60].

However, some of the analyses also reveal that these advantages are not without costs. Highlighting drawbacks of German bank-firm relationships, Albach (1998) suggests that it might be possible to overcome the lack of competition within the German financial system by strengthening the German capital market and by reducing the banks' influence on the supervisory boards of client firms.

Theory and empirical evidence seem to suggest that whether the advantages of bank-firm relationships outweigh the costs depends heavily on the economy's state of development, as well as the legal and institutional structure of its financial markets. Gerschenkron (1962) argues that initial stages of economic development can be better promoted by banking systems and controlled capital markets, whereas subsequent economic progress may benefit from more liberal securities markets. In an early analysis of Japanese growth between 1868-1940, Rosovsky (1961) applies Gerschenkron's framework to show the merits of developing close ties between manufacturing firms and financial institutions in order to overcome

---

verifiable by courts, whereas control-oriented finance only requires monitoring investors to be informed. On the other hand, the term "control-oriented" may suggest that there is no exercise of control under arm's-length finance, which is not true when we think of e.g. markets for corporate control. However, investor intervention is limited during the contracting period, and typically relies on third-party mechanisms that include the aforementioned market of corporate control as well as intervention via a court in case of bankruptcy.

59 German and Scandinavian data in particular.

60 It should be noted, however, that there are a few theorists like Blackwell and Santomero (1982), who do not agree that stronger bank-firm relations will always increase the firm's access to capital.

the "relative backwardness" of Japan's economy at that time. Only recently, an increasing number of scholars have started to argue that Japan held on to a highly regulated banking system and a controlled capital market for too long after it had already become a fully-developed economy.

Facing a most dramatic banking crisis and a troubled economy, the overwhelming majority of scholars as well as politicians seem to agree by now that, for the stage of development the Japanese economy has reached, the distortionary price for the provision of capital becomes a more troublesome source of inefficiency than during the early decades of high growth. Recent studies like that of Weinstein and Yafeh (1998) corroborate this view. They offer empirical evidence that "in the absence of contestable capital markets, large banks with close ties to industry siphon profits and restrict investment, and thus may inhibit rather than encourage growth" (Weinstein and Yafeh 1998: 666). Japanese authorities have subsequently recognized the need to adjust the financial system in order to secure continued economic progress. Beyond doubt, there has been a considerable time gap between recognizing the need for change and implementing necessary adjustments. Nevertheless, by now it is considered common knowledge that opening up arm's-length markets may make the most economic sense.

The following analysis is similar to the empirical studies cited above, to the extent that we examine both advantages and costs of bank-firm relationships. Our approach, however, differs from those of previous studies in that we emphasize differentiation between various ways of establishing and maintaining bank-firm relationships. Our unique data set allows us a closer look at whether corporate behavior can be influenced by the degree as well as nature of bank affiliations. It is possible to analyze ways in which close and continued interactions between banks and firms may provide firms with a competitive edge, and which aspects inhibit rather than encourage firm development. For this purpose, we do not want to restrict *a priori* the various possible interpretations. Hence, we refrain from using statistical selection processes that might eliminate specifications of bank-firm relationships from our regression models.

Most of the empirical analyses of Japanese bank-firm relationships have used data that include periods during which an underdeveloped Japanese capital market endowed financial institutions, especially banks, with a wide range of monopoly power. Hence, the impact of recent legal as well as institutional changes in Japan's financial markets could not, at least to its full extent, be captured by previous studies. Using a data set that covers the period between 1985-1998, an additional contribution of this chapter to the existing literature is that it sheds more light on how the dramatic process of deregulation and liberalization since the 1980s may have changed the economic role of Japanese bank-firm relationships.

The following analysis of the recent economic role of bank-firm relationships comprises a variety of input as well as output factors that concern the development of the firm. The rest of the analysis proceeds as follows: in chapter 5.2, we analyze whether bank affiliations increase the profitability of firms or are rather an obstacle to promoting profit. Chapter 5.3 observes the impact of bank-firm relationships on growth rates. We test whether bank affiliations enhance firm growth or are more likely to slow down further growth, especially of large, mature manufacturing firms. In chapter 5.4 we will look for empirical evidence that may support the view that bank-firm relationships are used to increase a firm's personnel welfare. The analysis proceeds with chapter 5.5 by analyzing whether bank-firm relationships may reduce firm risk. Chapter 5.6 discusses the impact of bank affiliations on the cost and stability of interest rates on borrowings. We first analyze whether bank affiliations may change the cost of bank loans. In a second step, we observe whether banks provide an insurance function by reducing interest rate fluctuations for client firms or rather offer incentives by monitoring and adjusting interest rates. The chapter closes with some concluding remarks in 5.7.

5.2 Profitability and Bank-Firm Relationship

5.2.1 Introduction

Set against firms without bank relationships, one of the often-cited competitive advantages of bank-affiliated Japanese firms is their easy access to capital (Aoki, Patrick, and Sheard 1994). Analyzing the impact of bank relationships on *keiretsu* affiliated firms, Hoshi, Kashyap and Scharfstein (1991: 49) find that "the closer a firm moves to the group banks, the more easily a firm can attract funds to finance investment projects", although they find no significant differences in investment levels. Note that analyzing U.S. data on small firms, Petersen and Rajan (1994) obtain similar results. In their study, they demonstrate that close relationships with a bank increase the availability of capital. However, bank relationships do not seem to reduce its costs.

The importance of bank-firm relationships in Japan becomes immediately clear when we reconsider the financial environment of corporate Japan until the 1980's. During most of the post-war period, firms could raise only limited amounts of capital through commercial floatation of debt or equity. Thus, firms mainly relied on internal and bank finance. In addition, Japanese capital markets were traditionally segmented, with the result that some firms had to cope with limited supply of capital while others enjoyed easy access to (bank) finance. Considering this environment, simple economic theory leads us to expect the bank-affiliated firms to outperform their unaffiliated, capital-rationed competitors.

However, as liberalization and with it the variety of financial sources other than bank debt have increasingly gained momentum since the early 1980's, the environment of corporate Japan concerning the supply of capital certainly has been changing. Chapter 4 already provided a model and empirical evidence to further illustrate the mechanism of bank-firm relationships, market value and debt. With financial liberalization facilitating access to a variety of outside financial sources, we note an increasing use of the capital market especially by highly valued firms, whether they are affiliated with banks or not. Being able to convince the capital market that it chooses the investment project with a high probability of success even without being forced by monitoring banks to act diligently, a firm does not necessarily need a bank relationship to secure easy access to outside capital. Since bank-affiliated firms have to compete increasingly with firms which can attract other sources of finance that are often cheaper than bank debt, we expect that the competitive advantage once claimed only by bank-affiliated firms now has to be shared with an increasing number of other firms. Firms that are highly valued by the capital market should be motivated to work profitably. Therefore a high market value should be positively correlated with profitability.

Even for the period between 1971-1982, however, when sources of outside finance other than bank finance were much more limited than in recent years, Nakatani's (1984) sample shows that despite easy access to capital, firms affiliated with a corporate financial group did not demonstrate higher rates of profit. He argues that the motivation for forming and maintaining a *kinyû keiretsu* must be found elsewhere than in profit maximization behavior. Using pooled data comprising the years 1977-1986, the OLS regression results of Weinstein and Yafeh (1998) confirm Nakatani's original findings for the case of main bank clients.

To summarize so far, we can state that bank-firm relationships have both advantages and disadvantages. The overall impact depends upon the prevailing type of bank-firm relationship and to what degree alternative sources of finance are available. In the following, we analyze whether the popular hypothesis that bank-affiliated firms outperform their unaffiliated peers is still valid for the boom-bust cycle. Note that we do not limit our analysis to the impact of main bank affiliation or *kinyû keiretsu* affiliation. Rather, we test whether the hypothesis holds true for the whole variety of bank-firm relationships introduced in chapter 3.2.

5.2.2 Empirical Analysis

In order to reflect the performance of relevant business activities for the period in question and to ensure that the measure is not subject to various artificial modifications which might not reveal the actual performance of the firms' activities, we follow Weinstein and Yafeh (1998), by measuring profitability as the rate of ordinary income over sales. In contrast to operating profits, this measure also includes receipts of interest and dividends as well as interest paid to financial institutions. It would be misleading to use only operating profits, since for many firms the receipts of interest contributed strongly[61] to the firm's overall profitability[62], especially during the bubble period.

The results of the fixed effects regressions, which are based on our panel data set that comprises the period between 1985-1998 (see chapter 3.1), are summarized in table 5-1. Controlling for further factors that may affect accounting measures of profitability like growth rates, leverage, etc. (see Weinstein and Yafeh 1995, 1998; Fisher 1987), table 5-1 underlines that depending on the nature of bank affiliation, bank-firm relationships have, if at all, a mixed effect on profitability.

Table 5-1: Profitability, 1985-1998

Dependent variable: ordinary income over sales
Fixed effects regression coefficients (T-values)

		(1)	(2)	(3)
Relationship Variables	KEILO	0.00101*** (3.559)	0.00099*** (3.548)	0.00096*** (3.456)
	DISPATCH	-0.00031 (-1.386)	-0.00037* (-1.687)	-0.00036* (-1.660)
	SHARE	-0.00028 (1.255)	-0.00039* (-1.792)	-0.00038* (-1.710)
	MB	-0.00277 (-0.861)	-0.00170 (-0.540)	-0.00253 (-0.805)
	Sig. relationship variables	0.000	0.000	0.000

Note: *, **, ***: Significance at the 10%, 5% and 1% level, respectively.

61 Schaede (1990) shows that using *zaiteku* ("high-tech financial management"), in the second half of the 1980s many manufacturing firms generated more than 50 per cent of their yearly profits from financial investment.

62 Running regressions with other measures of profits, e.g. business profits over total assets (see Nakatani 1984), produces similar results. However, in order to compare our results with those of Weinstein and Yafeh (1998), we focused on interpreting the regression using ordinary income over sales.

Table 5-1: Profitability, 1985-1998 (continued)

		(1)	(2)	(3)
Control Variables	Bank debt/ total debt	-0.1607*** (-4.304)	-0.01202*** (-3.292)	-0.01338*** (-3.642)
	MV	0.00452*** (6.624)	0.00566*** (9.494)	0.00481*** (7.194)
	Sales growth	0.04028*** (5.837)	0.04581*** (6.807)	0.04334*** (6.403)
	Debt/sales	-0.02598*** (-4.959)		-0.01484*** (-2.770)
	Log (sales)	0.05208*** (10.215)	0.04612*** (9.166)	0.04290*** (8.329)
	GDP	0.00648*** (11.484)	0.00553*** (9.752)	0.00557*** (9.845)
	Firm risk		-0.45118*** (-8.330)	-0.40603*** (-7.197)
	Size	0.00451* (1.891)	0.00303 (1.322)	0.00438* (1.874)
	Year dummies	Yes	Yes	Yes
Statistics	N of observations	1288	1288	1288
	N of groups	92	92	92
	Obs. per group	14	14	14
	Sign. F	0.000	0.000	0.000
	R^2	0.4758	0.4947	0.4980
	DW[63]	1.8450	1.8546	1.8583

Note: *, **, ***: Significance at the 10%, 5% and 1% level, respectively.

[63] For Durbin-Watson, here as well as for the analyses below, we followed Bhargava, Franzini, and Narendranathan (1982), who generalize Durbin-Watson type statistics to test the OLS residuals from the fixed effects model for serial independence.

To be more precise, the coefficient of the variable KEILO is positive and significant at the one per cent level. Similar to the analysis of Weinstein and Yafeh (1998), but unlike Nakatani (1984), the regression specifications in columns (2) and (3) include a measure of firm risk. We use the variance of the return on sales over five years (Geisen 1979)[64]. Hence, we can conclude that profitability differences between firms with a bank affiliation through *kinyû keiretsu* loan dependency and other firms cannot be due solely to differences in risk. Column (2) underlines that the results remain stable even when leverage is excluded from the regression (because it may be correlated with KEILO). Thus, other things being equal, a bank relationship that allows a strong usage of *kinyû keiretsu* loans appears to increase profitability.

However, the variables of all other measures of bank-firm relationships suggest a negative impact on profitability. The coefficients of the two variables DISPATCH and SHARE are both negative and in the regression specifications which include the measure for firm risk, they are both significant at the ten per cent level. This indicates that including a measure of risk in our specifications improves the explanatory power of our regression to describe the impact of bank-firm relationships on profitability.

The coefficient of the main bank dummy also shows a negative sign throughout all specifications. Even though the result is not significant at any acceptable level[65], it does to some extent corroborate the findings of Weinstein and Yafeh (1998), that a main bank affiliation does not seem to improve profitability.

Turning to the control variables, we find that profitability is a decreasing function of the bank debt to total debt ratio. The coefficient of the variable is negative and significant at the one per cent level. It remains unchanged throughout the regression specifications, indicating that the relatively low profitability of firms which use banks as an important source of capital remains unchanged even when firm risk is added to the regression or a measure for leverage is excluded. The results underscore that the difference between firms for which banks are an

---

64 Following Geisen (1979: 395), we define firm risk as the variance of the return on sales over five years at a time, so that

$$(1)\quad Risk_{i,t} = \sqrt{\sum_{m=0}^{4} \frac{\left(x_{i,t-m} - \overline{x_{i,t}}\right)^2}{5}};$$

with (2) $\overline{x_{i,t}} = \frac{1}{5}\sum_{m=0}^{4} x_{i,t-m}$ where x: return on sales.

65 Accounting for the number of firms under investigation, the level of significance should be ≤ 10 per cent (Kennedy 1985: 62; Leamer 1978: 88 and 104).

important source of capital and firms that use mostly alternative sources cannot be due exclusively to different degrees of firm risk.

Comparing the coefficient of the bank debt to total debt ratio with the coefficient of the variable KEILO, the result emphasizes that it is most important to differentiate between different sources of bank debt. While loans from financial institutions of the same *keiretsu,* which are most likely coordinated by the group's main bank, seem to help increase profitability, bank loans in general seem to do the opposite. Our interpretation of this finding is that when firms are allowed by their *keiretsu* banks to actually use *keiretsu* loans, the rents from this preferential access to capital do accrue largely to these firms. In other words, suppose the closeness to the group of banks of the same *keiretsu* can be described by the degree of access to *keiretsu* bank loans; it seems that once *keiretsu* affiliated firms have managed to establish a close relationship, they can actually enhance profitability. Hence, we cannot support the general statement made by Nakatani (1984) that firms affiliated with bank-centered corporate groups do not have higher rates of profit. Note, however, that in recent years not all *keiretsu* affiliated firms have had easy access to *keiretsu* loans (chapter 3.2.4.4). Taking this development into account, it rather seems to depend on whether or not firms affiliated with bank-centered groups manage to attract loans originating from banks of their group. Thus, the results do not suggest that a *keiretsu* affiliation itself is enough to enhance profitability.

Furthermore, the positive sign of the coefficient for firm size is quite surprising since it is widely believed that the rate of profits declines as firm size increases. Nevertheless, this finding is similar to the results of Nakatani (1984). Finally, as expected, profitability is positively correlated with GDP growth as well as with the ratio of market value to total liabilities.

5.2.3 Conclusion

To summarize, with regard to the significance of the vast majority of variables comprising different bank-firm relationships as well as accounting for the joint significance of these variables, we can observe that bank-firm relationships have an impact on the profitability of firms. However, depending on the nature of the relationship, the impact seems to differ in intensity as well as direction. While bank affiliations that lead to easy access to *keiretsu* loans enhance profitability, all other types of bank-firm relationships seem to have a negative impact on a firm's profit. Hence it becomes clear once again that our objection towards the main bank dummy or the dummy for *kinyû keiretsu* affiliation as being the only indicator for a bank affiliation is quite legitimate. It seems most important to differentiate between different bank-firm relationships in order to ascertain their advantages and disadvantages.

5.3 Growth Rates

5.3.1 Introduction

According to numerous analyses on corporate behavior, firms with close bank affiliations use their easy access to capital to advance their growth record rather than being primarily concerned about profitability, even in the long term (e.g. Hoshi, Kashyap and Scharfstein 1991; Beason and James 1999). Hence, one might assume that for bank-affiliated firms, size and therefore growth rather than profitability are most important. Indeed, in the case of the German bank-centered financial market system, Gerschenkron (1962) argues that large German banks have been able to continuously accelerate growth of their large corporate customers, while profitability was of secondary concern. Without doubt, the bank-centered financial system was of central importance for Japan in helping its economy develop and flourish from Meiji Restoration until the beginning of the Second World War (Rosovsky 1961; Yabushita and Inoue 1993), as well as during the rapid growth period between the mid-1950's to the mid-1970's (e.g. Aoki, Patrick, and Sheard 1994). These were the periods when firms were growing rapidly and were hungry for investment funds, without being much concerned about profitability. Especially during the post-war era of high growth, where financial markets were heavily regulated in favor of existing banks, strong bank-affiliations were certainly the cornerstone of corporate financial strategy and virtually essential for corporate success (Aoki, Patrick and Sheard 1994). At the end of this period, Japan advanced to become the world's second largest market economy, with European levels of per capita income.

Moreover, growth rather than profitability is often claimed to be of central interest for the employees of a Japanese firm, including the top management. To further understand this, note that under the Japanese seniority promotion system, faster growth leads to an increase in the probability of opportunities for faster promotions, hence more quickly raising salaries as well as status. Rapid growth furthermore strengthens the financing of the firm's pension scheme, since an employee of a fast growing firm will have many employees in that firm younger than himself (Komiya 1990).

However, analyzing the role of *keiretsu* affiliations, Nakatani (1984) finds no positive impact from *kinyû keiretsu* membership on the growth of firms. Furthermore, Weinstein and Yafeh (1998) show that annual growth rates of main bank clients are not much different from those of their peers. Considering the liberalization of the financial market, which gradually exposes Japan to a highly competitive environment where large firms especially have to compete for funds on an international level, it is hard to imagine that these firms can afford to be primarily interested in growth rather than safeguarding investors' interests.

5.3.2 Empirical Analysis

We have argued so far that the ability of a bank affiliation to positively influence a firm's profitability depends largely on the nature of that affiliation. In fact, only the variable for KEILO indicated that bank-firm relationships seem to improve firm profitability. It is therefore worthwhile to examine whether during times of liberalized financial markets, when the interests of investors are increasingly stressed, the potential advantages of bank affiliations are still revealed in faster growth rates rather than in profitability.

To measure the growth of firms, we used a wide range of criteria in initial modeling attempts, including the growth of sales, total assets, equity, gross performance and property compensation. In order to compare our results with those of Nakatani (1984), and Weinstein and Yafeh (1998), in the following we will analyze the regression results when using the growth rate of sales (table 5-2). However, the estimates based on the other measures of growth mentioned above are all very similar.

The joined significance of the relationship variables does not provide any evidence that bank-firm relationships have explanatory power for the firms' growth rates. Being negative and significant at the ten per cent level after including the measure for firm risk (column 2), though small, the coefficient of the variable DISPATCH indicates that growth rates of firms with bank dispatched directors are slightly lower than those of other firms. All other results for variables representing bank-firm relationships do not show significance at any acceptable level.

However, the coefficient of the bank debt to total debt ratio, being negative and significant at the one per cent level after controlling for the (log) level of firm sales lagged one period, underscores that annual growth rates of firms whose bank loans account for a high portion of total indebtedness are in fact lower than those of their peers during the sample period. It is furthermore interesting to note that, contrary to expectations, the growth rate of sales is positively correlated with the firm's size in terms of assets, whereas the coefficient of the risk variable suggests that a high level of firm risk leads to lower growth rates of sales.

Table 5-2: Growth rate of sales, 1985-1998

Dependent variable: Growth of sales compared to the previous year
Fixed effects regression coefficients (T-values)

		(1)	(2)
Relationship Variables	KEILO	-0.00009 (-0.080)	-0.00037 (-0.330)
	DISPATCH	-0.00132 (-1.475)	-0.00145* (-1.649)
	SHARE	0.00081 (0.904)	0.00046 (0.524)
	MB	0.00244 (0.191)	0.00111 (0.088)
	Sig. relationship variables	0.2218	0.2218
Control Variables	Bank debt/total debt	-0.04537*** (-3.080)	-0.03766*** (-2.597)
	Log (lagged sales)	-0.27265*** (-15.664)	-0.30468*** (-17.185)
	Firm risk		-1.42526*** (-6.761)
	Size	0.03635*** (3.958)	0.03895*** (4.317)
	Year dummies	Yes	Yes
Statistics	Number of observations	1288	1288
	Number of groups	92	92
	Obs. per group	14	14
	Sign. F	0.000	0.000
	R^2	0.4257	0.4472
	DW	1.7943	1.8143

Note: *, **, ***: Significance at the 10%, 5% and 1% level, respectively.

5.3.3 Conclusion

All in all, concerning the impact of bank-firm relationships on growth rates of firms, our results are close in line with those of Nakatani (1984) and Weinstein and Yafeh (1998). Claiming a close bank relationship when a firm is affiliated with a *kinyû keiretsu* or a main bank respectively, they do not find annual growth rates of affiliated firms to be larger than those of independent firms. Hence, we can conclude that at least since the 1980's, the often-claimed advantage of a bank relationship is not manifest in faster growth.

In fact, our results tentatively suggest that, if it has any impact at all, an affiliation with a bank may be an obstacle rather than a benefit to increasing annual growth rates. This result may corroborate the findings of Houston and James (1996). Though they focus on different aspects of bank-firm relationships, they find that since the early 1980s, firms that have maintained a long-term relationship with a single creditor were likely to have weak opportunities of growth. The result can be furthermore interpreted to suggest that bank-firm relationships are more common for mature firms with low growth rates.

5.4 Compensation of Employees

5.4.1 Introduction

Nakatani (1984) argues that rather than maximization of profit or growth, an appropriate motive for a *keiretsu* affiliation is the pursuit of employee interests. Obviously, Nakatani does not subscribe to the argument of Komiya (1990) mentioned above, that advancing growth rates support the interests of employees. Instead, he claims that the advantages of having easy access to capital via a close *kinyû keiretsu* affiliation benefit personnel welfare much more directly. Taking advantage of easy capital access in order to enhance employee welfare rather than increase profitability or growth would certainly be an indicator of the priority of employee interests over the interests of other stakeholders, especially those of shareholders. Furthermore, it would provide empirical evidence for the often-claimed intention of Japanese top management to value employee interests above shareholder interests (Gerlach 1992; Komiya 1990). It would also therefore be in line with the popular belief that the objective of Japanese corporations contrasts sharply with Anglo-Saxon corporate attitudes. In order to draw a more complete picture of the impact of bank-firm relationships on corporate behavior, we therefore continue our analysis by examining each of the possible forms of bank affiliations under consideration, and determining whether the relationship supports employee interests in the sense of directly enhancing personnel compensation.

5.4.2 Empirical Analysis

In our analysis of employee compensation, which focuses on the impact of bank affiliations, the average compensation to employees is computed as the ratio of the firm's total labor cost (including salary, other compensations, contributions to fringe benefits and social security) to the number of employees at the end of each accounting period[66]. Table 5-3 presents estimation results of the determinants of employee compensation.

Table 5-3: Average compensation to employee, 1985-1998

Dependent variable: ratio of the firm's total labor cost to the number of employees
Fixed effects regression coefficients (T-values)

		(1)	(2)	(3)
Relationship Variables	KEILO	-0.00516 (-0.742)	-0.00409 (-0.592)	-0.00463 (-0.671)
	DISPATCH	-0.00891* (-1.638)	-0.00894* (-1.657)	-0.00901* (-1.669)
	SHARE	-0.00183 (-0.335)	-0.00355 (-0.652)	-0.00329 (-0.603)
	MB	0.18995** (2.410)	0.17442** (2.227)	0.17709** (2.262)
	KEIRETSU		-0.15310 (-1.283)	
	Sig. relationship variables	0.0000	0.0000	0.0000

Note: *, **, ***: Significance at the 10%, 5% and 1% level, respectively.

66 Note that concerning the data on employees for our analysis, the *yûka shôken hôkokushô* discloses only data on core employees known as *sha-in* (company members), that is, permanent or long-term employees who "entered the firm" (*nyusha* or *shushoku*) upon graduation from school, although a few hired later could also be included. This group includes both blue-collar and white-collar workers. Part-timers, temporary employees, *shokutaku* (non-regular staff members) and "outside" workers (those hired and dispatched by a subcontracting company) are not included.

Table 5-3: Average compensation to employee, 1985-1998 (continued)

		(1)	(2)	(3)
Control Variables	Bank debt/total debt	-0.12362 (-1.347)	-0.10039 (-1.100)	-0.10745 (-1.179)
	ordinary income/sales	1.98068*** (4.267)	1.02737** (2.010)	1.03005** (2.015)
	capital/employees	0.00698*** (5.142)	0.00724*** (5.368)	0.00722*** (5.349)
	RME	1.20627** (2.140)	1.18362** (2.114)	1.16029** (2.073)
	Log (employees)	-0.43888*** (-2.727)	- 0.55683*** (-3.432)	-0.56078*** (-3.456)
	Firm risk		-6.02720*** (-4.258)	-6.07564*** (-4.292)
	Size	-0.02679 (-0.460)	-0.01008 (-0.174)	-0.01057 (-0.182)
	Year dummies	Yes	Yes	Yes
Statistics	Number of observations	1288	1288	1288
	Number of groups	92	92	92
	Obs. per group	14	14	14
	Sign. F	0.0000	0.0000	0.0000
	R^2 within	0.6643	0.6699	0.6695
	DW	1.9465	1.9548	1.9597

Note: *, **, ***: Significance at the 10%, 5% and 1% level, respectively.

As expected, the average compensation to employees is positively correlated with profit (income over sales), hence suggesting the existence of "profit sharing" between employees and other stakeholders in Japanese firms (Nakatani 1984; Komiya 1990). The bonus system, where the amount paid to employees varies according to the firm's success, is one example for the channels frequently used to allocate additional payments so that employees participate directly in the firm's profit (loss). Though not significant, the coefficient of the bank debt to total debt ratio tentatively suggests that the more dependent a firm is on rather costly bank loans as a source of outside finance, the smaller is its ability to pay high compensations for employees. Explanatory variables furthermore include measures like capital intensity that may affect accounting measures of employee compensation. It is no surprise that with an increasing ratio of capital per employee, employee compensation increases. Note that the coefficient of firm size in terms of assets has a negative sign but shows no statistical significance at any acceptable level, which suggests that for the firms included in our analysis, employee compensation is not a function of firm size. It is furthermore interesting to note that employee compensation is a decreasing function of firm risk.

Following Hadley (1970) and Nakatani (1984), we also include a ratio that reflects the age structure as well as the male-female composition of employees (RME). This is the ratio of male employees to total employees adjusted with respect to the age structure of the work force of each firm[67]. As shown in table 5-3, the coefficient of RME is positive and significant at the five per cent level. The positive correlation between the average compensation to employees and the age-corrected ratio of the male-female composition of employees shows that for the sample analyzed, the seniority wage system and the fact that male employees earn more than their female colleagues is still vivid. Thus, males counting among the core employees (*sha-in*) of large Japanese firms seem to be still largely protected by the umbrella of the lifetime employment and the seniority promotion system. In this respect, Nakatani's (1984) findings from his 1971-1982 sample are corroborated by our 1985-1998 sample results.

For the purpose of this analysis, however, the most important coefficients are those of the bank-firm relationship variables. Testing for the joint significance of our bank-firm relationship variables, we find that bank affiliations have an impact on the average compensation of employees. Nevertheless, the impact is heterogeneous in intensity as well as direction: The coefficient of the main bank dummy is positive and significant at the five per cent level. However, although the statistical significance is not high for most of the other relationship variables, the fact that they are all negative suggests that bank-firm relationships other than a main bank affiliation do not raise the average compensation of employees. Notice that the regression specification presented in column (3) includes (unlike Nakatani's (1984) experiment) the measure of firm risk, which has already been

67 The ratio of male employees to total employees adjusted with respect to the age structure of a firm's work force has been computed as follows:

$$\text{RME} = \frac{ME \times AGM}{ME \times AGM + FE \times AGF}$$

with ME: number of male employees; AGM: average age of male employees; FE: number of female employees; AGF: average age of female employees (see Hadley 1970).

used above. Hence we can conclude that differences of employee compensation between firms with bank relationships and other firms cannot be due solely to differences in risk.

5.4.3 Conclusion

To summarize, only main bank affiliations have a positive impact on the compensation of employees, while all other types of bank-firm relationships under observation tend to have a negative influence. Our interpretation of this finding is that rents which may be due to bank dispatched directors, a high percentage of client firm share held by banks, or a strong use of *keiretsu* loans, do not enhance employee welfare at banks' client firms. To double check for any result that might support Nakatani's (1984) argument, we furthermore run a regression specification that includes a dummy for *keiretsu* affiliation called KEIRETSU. Column (2) of table 5-3 presents the estimation result. The coefficient of the *keiretsu* dummy shows a negative sign without being significant at any acceptable level. Thus, we cannot corroborate the findings of Nakatani (1984).

Concerning main bank relationships, neither profit maximization nor growth maximization was shown to be an appropriate motive for a main bank affiliation. However, the positive and highly significant coefficient of the main bank dummy suggests that, controlling for other factors, employees of firms with a main bank are better compensated than those of firms without a main bank affiliation. The results suggest that rather than increasing profitability or growth, gains from main bank affiliations may be used to improve employee welfare. We do not have a straightforward explanation for this result. However, one of the reasons why only main bank affiliations seem to have a positive influence on the level of employee compensation may be their special position concerning the management of employee-related services. In order to maintain this lucrative position, the main bank may use its influence on the behavior of the firm to provide employee compensation. Recalling furthermore the benefits that accrue to a bank once it has become a firm's main bank (see chapter 3.2.4.4), a bank-firm relationship based on a main bank affiliation truly mutually benefits firm employees and the main bank.

5.5 Firm Risk

5.5.1 Introduction

Despite advantages of access to capital, bank-affiliated firms apparently do not have any significant advantage over their peers when it comes to growth rates (chapter 5.3). Trying to explain these circumstances, Weinstein and Yafeh (1998) suggest that in comparison with independent firms, banks may induce their client firms to be more risk averse when deciding about investment projects. Their finding is close in line with Miarka and Yang (1997), who show that in comparison to their peers, large Japanese manufacturing firms with a strong dependency on bank loans decide on more risk-averse investment projects. Furthermore, Montalvo and Yafeh (1994) claim that, having a voice in the firm's affairs, a main bank influences its client to decide on risk-averse ways to access new technologies. In their study they show that rather than acquiring knowledge of new technologies by developing them in-house, which would make risky investment in R&D more likely, main bank clients are more inclined to gain access to new technologies via licensing of relatively safe, proven foreign technologies. Corroborating this finding, Yafeh and Yosha (1999) show that bank-affiliated firms tend to spend less on R&D than unaffiliated firms. These observations, which all demonstrate that bank affiliations reduce firm risk, are also close in spirit with our findings discussed in chapter 4. We showed that close bank-firm relationships based on bank-dispatched directors (DISPATCH), as well as relationships based on a large percentage of client firm shares held by banks (SHARE) lead to increased monitoring, which decreases the private benefits of risky behavior. Hence, when analyzing the impact of bank-firm relationships on firm risk, we expect bank affiliation to reduce firm risk.

5.5.2 Empirical Analysis

Table 5-4 presents the estimation results of the determinants of firm risk. For the measure of firm risk, we continue to use the variance of the return on sales over five years at a time (see chapter 5.2)[68]. MV is, as before (chapter 4.3), the firm's market value compared to its total liabilities. It is positively correlated with firm risk, which is consistent with basic investment theory: investments with higher risk should only be made if the risk is compensated by a higher expected rate of return. Furthermore, firm risk is negatively correlated with profitability, which we have measured once again as ordinary income over sales. This is consistent with our earlier finding showing that profitability is negatively

---

[68] Note, however, that further modeling attempts showed similar results when using the variance of the return on capital.

correlated with firm risk (see chapter 5.2, Table 5-1). Note that in order to control for factors that may affect accounting measures of our risk variable, the regression specification furthermore includes the measurements debt over sales and log (sales). Moreover, we include a measure that controls for the general development of the Japanese economy during the boom-bust cycle: the coefficient of the annual growth rate of Japan's gross domestic product in real terms shows a negative sign and is significant at the one per cent level. Thus, as expected, risk increases with decreasing GDP growth.

More importantly, let us turn to the results of the bank-firm relationship variables. Similar to regression specifications that have been carried out in initial modeling attempts concerning bank-firm relationships and bank debt (chapter 4.3), the coefficient of the main bank dummy as well as the coefficient of the variable representing the dependency on bank loans originating from the affiliated *kinyû keiretsu* (KEILO) are not significant at any acceptable level.

However, recalling the analysis presented in chapter 4, the coefficients in which we are most interested are those of the variables DISPATCH and SHARE, as well as of the ratio of bank debt to total debt. The variables accounting for bank debt and the variables SHARE and DISPATCH are significant at the one, five, and ten per cent level, respectively. Observing the direction of the three variables, as expected, the coefficients of the two bank-firm relationship variables have a negative sign while the coefficient for bank debt is positive. Concerning the results for the bank-firm relationship variables, it allows us to assert that, controlling for other factors, close bank-firm relationships established by bank dispatched directors (DISPATCH) or based on a large percentage of client firm shares held by banks (SHARE), reduce firm risk. Given our model, the result of the bank debt ratio is not surprising either: In the model, we explained that firms have to overcome moral hazard problems in order to have easy access to (cheaper) capital market finance. A firm with a high need for finance compared to its market value may overcome the moral hazard problem by inciting banks to monitor its actions. By doing so, the firm may convince the dispersed investors in a competitive market that it behaves diligently. However, since monitoring is costly for the bank, the firm needs to ask for a sufficiently high amount of bank loans (which are costly for the firm) in order to be monitored adequately. We furthermore showed that many bank-firm relationships have mainly historical origins and are therefore exogenous (see chapters 3 and 4), reducing monitoring costs and therefore the amount of bank loans that are needed to incite the bank to monitor sufficiently. Nevertheless, as risk moral hazard is more difficult and therefore more costly to monitor in risky firms, everything else being equal, they have to take on a higher amount of bank loans than their peers. Hence, firm risk has to be correlated positively to the number of loans originating from banks.

Table 5-4: Firm risk, 1985-1998

Dependent variable: variance of the return on sales over five years at a time; Fixed effects regression coefficients (T-values)

		(1)
Relationship Variables	KEILO	-0.00001 (-0.054)
	DISPATCH	-0.00210* (-1.774)
	SHARE	-0.00270** (-2.442)
	MB	0.003550 (0.222)
	Sig. relationship variables	0.0000
Control Variables	Bank debt/total debt	0.00497*** (2.653)
	MV	0.00114*** (3.307)
	Ordinary income/sales	-0.09604*** (-6.707)
	Debt/sales	0.02390*** (9.346)
	Log(sales)	-0.01601*** (-6.336)
	GDP	-0.00139*** (-4.866)
	Year dummies	Yes
Statistics	Number of observations	1288
	Number of groups	92
	Obs. per group	14
	Sign. F	0.0000
	R^2	0.3138
	DW	1.9452

Note: *, **, ***: Significance at the 10%, 5% and 1% level, respectively.

5.5.3 Conclusion

Summing up, the findings of our analysis on bank-firm relationship, bank debt and firm risk fully support the results of our previous study on bank-firm relationships and bank debt, discussed in chapter 4. Furthermore, the results confirm the findings mentioned above, which all underline that bank affiliations may reduce firm risk. However, though being generally similar in spirit to Weinstein and Yafeh (1998), we are not able to corroborate their conclusion that a main bank affiliation is the particular form of bank relationship which induces the firm to decide for more risk-averse strategies. Our findings suggest rather that bank-firm relationships based on dispatched directors and a high percentage of bank-held shares are most effective at reducing risk.

5.6 Interest Rates on Borrowings: Cost and Stability

We now turn to the issue of the costs of borrowing as well as the stability of these costs for bank-affiliated firms versus independent firms. We continue to focus on large firms in times of increasingly liberalized and deregulated financial markets.

5.6.1 The Cost of Bank Loans

5.6.1.1 Theories and Previous Empirical Results

For large, reputable firms like those in our sample, the relative cost of bank borrowing is likely to be higher than the cost of borrowing from capital markets (see Diamond 1991; Rajan 1992). Since the bank monitors the firm and controls its investment decisions, it is in the position to alter the division of surplus between itself and the firm. Similar to Rajan's (1992) analysis on the consequences of ex post rent extraction, Sharpe (1990) also assumes that banks can extract rents from their closely affiliated client firms, because they have better information about the borrower's credit risk. However, Tröge (1999) shows that this effect may disappear if outsiders can observe the interest rates offered by the insider. He argues that the advantage of inside banks results from their ability to provide better corporate governance rather than their better knowledge of the client firm's quality.

No matter what the reasons for the advantage of inside banks and thus their ability to extract rents, we can generally agree with Rajan (1992) that especially for large, reputable firms, there is a fundamental trade-off between bank finance and arm's-length finance. With increasing choices between various sources of finance, this trade-off becomes more and more apparent. Examining the impact of competition on the cost of capital, in an extension of Diamond (1989), Petersen and Rajan (1995) suggest that in contrast to small firms, where a bank's competition comes mainly from other banks, the main source of competition for banks to fulfill the financing requirements of larger firms comes from arm's-length sources like bond markets. When competition is from arm's-length markets rather than from other banks, it seems even more difficult for banks to retain the business of successful firms. Hence, theory suggests that for banks, one possible consequence of the further liberalization of arm's-length credit markets is to have to lower the costs of bank borrowings in order to stay competitive.

Surveying the literature on empirical analyses of Japanese bank-firm relationships, it is widely claimed that banks are able to extract significant rents from their closely affiliated client firms (see e.g. Weinstein and Yafeh 1998).

However, analyses of Japanese bank-firm relationships also strongly underscore the benefits a firm may gain in return for higher interest rates. Credit from relationship banks goes along with a wide range of advantages that cannot be obtained from other sources of credit, e.g. arm's-length sources. As we have already pointed out in chapter 4, a sufficient fraction of bank debt can incite the bank to monitor the firm appropriately, which decreases the entrepreneurs' incentives for risky behavior and therefore increases the firm's access to additional sources of finance. Furthermore, having stable bank affiliations, many firms have been more or less able to count on their banks' willingness to lend to them. As has been pointed out by Horiuchi (1994), especially in the case of routine investment, a substantial number of large Japanese firms simply inform their banks of the investment plans after company managers have already decided them. Also, it is claimed that in times of financial distress, bank-affiliated firms can generally count on the main bank as a lender of last resort and a source of competent managerial guidance. According to most empirical analyses (e.g. Aoki and Patrick 1994), this system of "give and take" has been relatively stable throughout the post-war period. However, we must note that most of the empirical analyses that support this view examine the advantages and costs of bank-firm relationships in a period where Japanese financial markets were highly regulated and immature. At that time, the financial system largely favored already-existing banks, and competition from alternative sources of outside finance was limited (see chapter 2).

With increasing financial liberalization on the one hand, and the position of banks as the major source of outside finance deteriorating since the 1980s on the other, easy access to financial sources seems to be an issue of declining importance, especially for highly valued firms. As mentioned above, the research results reported by Weinstein and Yafeh (1998) support the hypothesis that banks can extract higher rents from their affiliated firms. However, splitting up their observation period, they find evidence that the premium extracted by banks in return for improved access to capital declined during the later years of their inquiry, during which effects of the financial market liberalization and deregulation gathered momentum. Their results suggest that the premium banks could once easily extract from their dependent client firms via interest rates has been under pressure since their position as the only major source of outside finance began deteriorating.

Following Rajan (1992), a possible explanation for this development could be as follows: Suppose that a firm has to finance an investment project with external sources in order to obtain a stochastic payoff. In the case of a positive net present value, during the pre-deregulation period when the firm's source of finance was basically limited to bank debt, the firm may have shared a substantial portion of the surplus from the project with the banks in order to persuade them to continue lending. However, with the firm's freedom to diversify its sources of finance since the 1980's, the banks' ability to influence the division of surplus seems to have declined. As has been already pointed out in chapter 4, especially highly valued firms are usually able to easily obtain credit from sources other than banks. This leads to a diversification of financial sources, and thus to a decreasing capacity for the banks to extract high interest rates. In fact, Weinstein and Yafeh (1998), providing empirical evidence suggesting that banks seem to charge profitable clients less than unprofitable ones, are much in line with Rajan (1992), who argues that banks' ability to control and to influence the division of surplus are linked. In addition, evidence provided by Hoshi, Kashyap and Scharfstein (1990) shows that in the wake of increasing competition with arm's-length credit markets, firms with relatively high growth and high Tobin's q reduce their bank ties and borrow from capital markets.

5.6.1.2 Empirical Analysis

In our empirical analysis we examine to what extent bank affiliations (still) had an impact on the cost of bank loans during the boom-bust period between 1985-1998. We then proceed in 5.6.2 by analyzing the influence of differing types bank-firm relationships on the stability of interest rates over the same period[69].

When analyzing the cost of bank loans from the firm's point of view, ideally

---

[69] Note that in contrast to Petersen and Rajan (1994, 1995) or Berger and Udell (1995), we do not use the duration of a bank-borrower relationship as a measure of its strength, but rather continue to utilize the same measures of bank-firm relationships as have been already used above. In their analysis, Berger and Udell (1995) impose a maximum limit of 30 years on the variables that account for the age of the firm as well as the continuous duration of a relationship. They claim that no additional relevant information is revealed after 30 years. Taking into account that the youngest firm in our sample was founded in 1970 and most Japanese relationships between large manufacturing firms and banks, if any, were established in the 1950s (many even in the pre-World War Two period), duration would not be an appropriate measure for analyzing large, reputable Japanese firms.

the actual interest rates on bank loans should be examined. Unfortunately, such data are not available. Instead, we follow Weinstein and Yafeh (1998) and use the average interest rate on nonbond liabilities, that is, interest payments and discount expenses divided by nonbond current and fixed liabilities. Though this measure includes some components in the denominator that do not bear interest, it is virtually impossible to correct for this due to the sometimes insufficient division of balance sheet positions. However, unlike the Weinstein and Yafeh study where the interest rate itself is taken as the dependent variable, we follow Berger and Udell (1995) in that we furthermore correct for underlying prime rates by considering the difference between the yearly average of a firm's interest rate and the average prime rate per year[70]. For our regression, we once again use a two-sided fixed effects model. In the model, the cost of borrowing is a function of variables describing different kinds of bank-firm relationships and control variables, including variables that account for firm characteristics.

Based on various theoretical rationales, our general expectations regarding the signs of the fixed effect coefficients are as follows: First, firms will have to pay higher interest rates as leverage increases. The same ought to be true for a high dependency on bank-loans. The firm-risk variable ought to carry positive coefficients since lenders are likely to increase demand in the face of a high-risk firm. We suspect that more successful firms with greater bargaining power will be relatively more successful than less prosperous firms in avoiding high interest rates. It is equally plausible to us that this is also true for larger firms when compared to smaller firms.

As to the relationship variables, it is more difficult to hypothesize since the Japanese market has been changing a great deal in recent years. Theory (e.g. Diamond 1984, 1991) as well as recent empirical evidence (e.g. Weinstein and Yafeh 1998) underline that relationships lower the lender's cost of lending to client firms. However, it is far from clear to what degree, if at all, the lender is willing to pass along these positive effects to the borrower in the form of lower loan interest rates. As mentioned above, Petersen and Rajan (1994) note that whether the banks are forced to pass on the generated cost savings depends much upon the competitiveness of the capital market. The Weinstein and Yafeh (1998) study tentatively suggests a decreasing trend in the interest rate bank-affiliated firms had to pay in the first half of the 1980's. If this development is a result of deregulation and increasing competition between arm's-length and bank finance, we believe that this trend will have progressed much further during our observation period. However, though competition might further reduce the interest rates for bank borrowing, we do not necessarily expect that banks will pass on the savings. Supposing that the information generated by bank-firm relationships is (at

---

70 Prime rates change dramatically over the observation period, peaking in 1985 with 5 per cent and reaching a low of 0,5 per cent in September 1995, which stayed stable until the end of the observation period (Economic and Financial Data on CD-ROM, Bank of Japan 1999).

least to some extent) private to the banks and not completely transferable by the borrower to others, relationship banks may reduce the interest rate by less than the true decline in cost (Greenbaum, Kanatas, and Venezia 1989; Sharpe 1990; Rajan 1992; Petersen and Rajan 1994).

The results from our fixed effects specifications are summarized in table 5-5. Similar to Harhoff and Körting (1998b), we introduce the exogenous variables in groups in order to observe how the correlation between some of them affects the results. Before turning to the role of relationships, it is important not only to take account of the underlying prime rate, but also to control for firm-specific characteristics that may influence the interest rate. Hence, we start with a model that only includes firm-specific characteristics and control variables.

Looking at the regression results presented in column (1), we find the following: As expected, interest rates are an increasing function of the leverage ratio. The coefficient of the variable indicating the dependency on bank loans as the major source of finance is also significant at the one per cent level and shows a positive sign. Hence, our result gives some evidence that banks can extract higher interest rates, especially from firms that continue, despite financial liberalization, to be strongly dependent on bank finance rather than diversify their financial sources. The variable indicating firm risk is also positive, thus somewhat indicating that with increasing firm risk, the lender tends to ask for a higher interest rate. Larger firms seem to pay lower interest rates, possibly because they are perceived to be less risky. Note, however, that both the variable accounting for firm risk and the one representing size are not significant at an acceptable level. One can see that fast growing firms (in terms of sales growth as well as employment growth) seem to face lower interest rates than those with high returns on sales. Though showing a negative sign as expected, to our surprise the coefficient of the variable for the firm's market value over total debt is not significant at an acceptable level. Furthermore, we find that the coefficient estimates we obtain for the variable indicating the propensity to investment as well as for the variable accounting for the firm's ability to pay back its debt (profit/interest), are both negative and significant at the one per cent level. This suggests that fast growing firms which are willing to expend their investment, as well as firms which have no problem meeting their debt obligations, certainly have some bargaining power when it comes to interest rates on bank loans. We suppose that these firms are quite capable of using a large variety of financial sources other than bank debt. As a result, they have to pay lower interest rates on their liabilities than their peers.

Table 5-5: Interest rates on borrowings, 1985-1998

Dependent variable: average interest rate on borrowings
Fixed effects regression coefficients (T-values)

		(1)	(2)
Relationship Variables	KEILO		-0.01448** (-1.919)
	DISPATCH		0.01209** (2.030)
	SHARE		-0.00660 (-1.107)
	MB		-0.22267*** (-2.619)
	Sig. relationship variables		0.000
Control Variables	Bank debt/total debt	0.57583*** (5.869)	0.63586*** (6.396)
	MV	-0.01002 (-0.596)	-0.01234 (-0.737)
	ordinary income/sales	1.04044** (1.737)	1.09933** (1.836)
	Leverage	0.15900*** (5.618)	0.15188*** (5.352)
	Sales growth	-0.68135*** (-3.693)	-0.67139*** (-3.653)
	Propensity to invest	-0.39703*** (-5.017)	-0.39104*** (-4.958)
	Profit/interest	-0.01308*** (-6.923)	-0.01301*** (-6.901)
	employment growth	-0.37859 (-1.085)	-0.46271 (-1.325)

Note: *, **, ***: Significance at the 10%, 5% and 1% level, respectively.

Table 5-5: Interest rates on borrowings, 1985-1998 (continued)

		(1)	(2)
Control Variables	Firm risk	2.08686 (1.345)	1.99229 (1.285)
	Size	-0.09056 (-1.471)	-0.09394 (-1.529)
	Year dummies	Yes	Yes
Statistics	N of observations	1288	1288
	N of groups	92	92
	Obs. per group	14	14
	Sign. F	0.000	0.000
	R^2	0.8597	0.8614
	DW	1.9572	1.9738

Note: *, **, ***: Significance at the 10%, 5% and 1% level, respectively.

For the purpose of our analysis, however, the most important coefficients are those relating to bank affiliations. In column (2) we include the relationship variables among the regressors[71]. Comparing column (1) and (2), the exclusion of the relationship variables has virtually no effect on the coefficients in column (1). As for the variables indicating bank-firm relationships, the measure of joint significance already suggests that bank-firm relationships play a significant role in the pricing of loans. Asking for bank loans from banks of the firm's own *keiretsu* (KEILO) seems to enable the firm to access lower interest rates. As has been already mentioned above, Weinstein and Yafeh (1998) showed that interest rates for firms with a main bank affiliation decrease while liberalization and deregulation are gaining ground. Controlling for other factors, the coefficient of the main bank dummy in our specification underlines that this trend seems to have progressed substantially after the mid-1980s. In fact, given a coefficient that is negative and significant at the one per cent level, a main bank affiliation can be assumed to lower interest rates during the second half of the 1980s and most of the 1990s. Although statistical significance is not high, the coefficient for the variable accounting for the percentage of shares held by banks carries a negative sign as well.

---

[71] As in previous analyses, we also tested for possible nonlinear relations: however, in none of the analyses could nonlinear relations be found.

However, the coefficient on the influence of dispatched bank representatives is significant and positive. One of the reasons for this result might be that sending a highly capable bank representative to the firm is exceptionally costly for the bank. Recall that bankers who take a high position in the firm's board of directors are especially often dispatched in order to carry out consulting and management assignments in order to improve the firm's position. As was discussed in chapter 3.2.4.1, it is also generally agreed that having a bank representative in a high-ranking position on the board increases credibility. Apparently, despite increasing competition from arm's-length finance, banks are still able to ask for the compensation of such services (which arm's-length finance cannot offer) via interest rates. In fact, our result corroborates reported evidence (Bank of Japan 1992) that firms which have chosen to maintain close bank-affiliations via dispatched directors can keep the banks' liquidity services and "bankruptcy insurance scheme" in exchange for a liquidity and bankruptcy insurance premium.

5.6.2 Interest Rate Fluctuations

5.6.2.1 Theories and Previous Empirical Results

Besides easing access to capital, Japanese bank-firm relationships are alleged to protect closely affiliated firms from various fluctuations in the cost of capital that would make investment decisions difficult. In a series of articles, Hoshi, Kashyap and Scharfstein (1990a, 1990b, 1991) show that Japanese firms closely affiliated with a main bank are less vulnerable to fluctuations in cash flow and have lower costs of financial distress than independent firms. Following the implicit long-term contract theory, under the assumption that the bank is risk neutral and the firm is risk-averse, it is to the benefit of both parties to fix the rate of interest over time (Teranishi, Goto and Serizawa 1975). In fact, for his 1971-1982 sample, Nakatani (1984) shows that for *keiretsu* firms which are affiliated with at least one *keiretsu* bank, the lending rate is more stable over time, thus providing evidence that supports the hypothesis of the implicit long-term contract theory.

Nevertheless, since liberalization in the 1980s and especially with the burst of the speculative bubble in the early 1990's, risks as well as competition among the various sources of debt have become larger and banks weaker. As banks may have been disciplined by this process to offer particularly better-performing customers tailor-made, "competitive" interest rates in order to be more competitive against other sources of finance, it is not clear anymore whether they can continue to keep the interest rates for their closely affiliated customers stable over time. In fact, anecdotal evidence suggests that banks increasingly have to make use of the "Material Adverse Change" clause[72] most credit contracts contain, thus often changing (increasing) interest rates and sometimes calling back credit[73]. This raises some doubt as to whether in a time of increasing competition and an overall weakened banking sector, banks are still able to provide the insurance scheme against the fluctuation of interest rates to the same extent as has been observed in earlier years. Also note that other insurance schemes like averting bankruptcy seem to erode as well (Financial Times 28.10.1998: 16). While the benefits of bank debt seem to decrease, a rising number of firms may prefer credit from arm's-length sources, which neither provides the benefits of bank debt nor incurs the costs.

Clearly, if creditors rely on interest rates as their only or at least primary source of revenues, and given that the creditors have to compete in a highly deregulated market, it seems hard to smoothen interest rates since abilities to spread interests over time or to compensate for lower-than-competitive interest rates with other revenues are few. The analysis by Petersen and Rajan (1995) already suggests that in a competitive environment, despite long-term relationships, it is difficult for creditors to charge a lower-than-competitive interest rate at any point in time over the life cycle of the firm. They emphasize that competition may continuously prevent the creditor from charging a higher-than-competitive rate that could compensate/anticipate the loss that will be or has been incurred by offering a lower-than-competitive interest rate. In other words, with increasing competition, the ability of the creditor to obtain surplus from a firm during one period, which could be used to offset losses obtained during another period is weakened. Consequently, it is difficult to accept lower returns up front.

5.6.2.2 Empirical Analysis

In order to observe the stability of interest rates, as our dependent variable we once again take the difference between the average interest rate and the prime rate and then compute its variance over five years at a time for every sample firm and year between 1985-1998[74]. The variance is computed similar to the way we

72 Whereas the interest rate and the duration of a bond are fixed once for all, most credit contracts contain an unspecific "Material Adverse Change" clause (MAC) which gives the bank the discretion to call back the credit or increase the interest rate whenever this seems appropriate (Petersen and Rajan 1995; Tröge 1999). Though banks had been reluctant to invoke this clause in earlier years, the extreme conditions especially after the burst of the bubble forced the banks to invoke it more often (KPMG Peat Marwick 1993).

73 Interview with an anonymous director of Daiwa Shôken, London, 23.07.1997.

74 Note that, of course, for constructing the variance of the difference between the interest rate and the prime rate over five years at a time between 1985-1998, in a first step we had to construct the difference between the interest rate and the prime rate for the period between 1981-1998.

computed the variance of the return on sales (see chapter 3.5). Hence, unlike Nakatani (1984), and Wakita (1983), we focus on the variance of the interest on nonbond liabilities rather than on all interest bearing liabilities, and we correct for underlying prime rates.

Table 5-6 presents estimation results of what determines our dependent variable. Explanatory variables include the same measures for variables describing different ways of bank-firm relationships and control variables, as had been used for our estimation on interest rates. Results of the control variables are as expected: The coefficients of variables like the firm's market value or the ratio of ordinary income over sales are all negative and significant. Hence, the variance of the difference between the interest rate and the prime rate is a decreasing function of measures that account for the quality of the firm. With increasing dependence on bank loans, the dependent variable also increases.

For the purpose of this chapter, however, the coefficients of our bank-firm relationship variables are more important: The dominant determinant of the variance of interest rates appears to be whether or not a firm maintains a main bank affiliation with a coefficient of -0.23 (standard error 0.038; significant at the one per cent level). Apparently, the lending rate of interest is more stable for firms affiliated with a main bank. This result is very much in line with Nakatani (1984: 242), who could show that „the lending rate of interest is made more stable over time for the firm involved in a *keiretsu* relationship with a bank."

Table 5-6: Stability of the interest rate, 1985-1998

Dependent variable: variance of difference between interest rate and prime rate
Fixed effects regression coefficients (T-values)

		(1)
Relationship Variables	KEILO	0.00210 (0.627)
	DISPATCH	0.00879*** (3.334)
	SHARE	0.00684*** (2.589)
	MB	-0.22900*** (-6.081)
	Sig. relationship variables	0.000

Note: *, **, ***: Significance at the 10%, 5% and 1% level, respectively.

Table 5-6: Stability of the interest rate, 1985-1998 (continued)

		(1)
Control Variables	Bank debt/ total debt	0.14332*** (3.254)
	MV	-0.02582*** (-3.481)
	ordinary income/sales	-0.46469** (-1.752)
	Leverage	-0.04965*** (-3.950)
	Sales growth	-0.10003 (-1.229)
	Propensity to invest	-0.08163** (-2.337)
	Profit/interest	-0.00002 (-0.020)
	employment growth	0.19605 (1.267)
	Firm risk	-0.54778 (-0.798)
	Size	0.10455*** (3.842)
	Year dummies	Yes
Statistics	N of observations	1288
	N of groups	92
	Obs. per group	14
	Sign. F	0.000
	R^2	0.7364
	DW	1.9798

Note: *, **, ***: Significance at the 10%, 5% and 1% level, respectively.

However, combining this with our previous findings on the effect of a main bank affiliation on interest rates, we find the result quite surprising. Seemingly, main banks are able to reduce the risk of interest rate fluctuations for their clients without asking for higher interest rates from them. Our interpretation is that being the firm's main bank seems to be so valuable for the bank that it keeps on providing services, like protecting the client firm from the risk of strongly fluctuating interest rates, in order not to loose its overall highly profitable position. It seems plausible to us that in times where lending rates diminish in profitability, the main bank position, which includes attending to the firm's employee accounts as well as lucrative fee-based and commission banking (chapter 3), is of exceptional importance for banks (Scher 1997). Putting it differently, in contrast to other banks, the main bank manages much better to extract compensation for its services from privileges that are not directly related to the provision of loans. In contrast to the provision of capital, where competition increased dramatically, main banks have obviously been able to defend other areas of banking where long term trust, inside information and firm-specific competence are more difficult to obtain. The rents which main banks can extract from these services seem to be high enough to allow for continued smoothening of interest rates, despite increasing competition in the credit market. Hence, in contrast to positions of other banks, a main bank position that insures access to lucrative services other than the provision of loans makes it much easier to spread the costs that might occur e.g. in the smoothening interest rates, over multiple products.

In fact, the coefficients of the other three relationship variables seem to further support this view: all of them describe different forms of bank-firm relationships which do not necessarily go along with the lucrative services a main bank is most likely allowed to offer. Therefore, these types of bank affiliation do not seem to have the ability to provide the same insurance scheme a main bank manages to offer. Hence, it is no surprise that none of the coefficients of the remaining three bank-firm relationship variables carry a positive sign. To be more precise, the results suggest that the types of bank-firm relationships represented by the variables KEILO and SHARE would neither allow the banks to fully remunerate potential losses through lucrative services as discussed above, nor would banks involved in these types of affiliations necessarily manage to compensate losses by extracting higher-than-competitive interest rates. As to the type of bank-firm relationship represented by DISPATCH, the results show that banks which have dispatched representatives to the boards use the revenue from higher-than-competitive interest rates in order to compensate the cost of dispatching (see chapter 3.6.1), rather than to protect clients from interest rate fluctuations. To summarize, our findings are similar in spirit to Petersen and Rajan (1995): creditors in competitive markets, which are not compensated for their services and not able to spread possible costs that might be caused by the smoothening of interest rates over multiple products, have no assurance anymore of obtaining future surplus from the firm via interest rates. Thus they are unable to secure their clients against the risk of interest rate volatility.

5.6.3 Conclusion

Our analysis highlights the impact of bank-firm relationships on the cost of commercial bank loans and the stability of these costs after accounting for underlying prime rates. Once again, the evidence indicates that it is necessary to distinguish between the different forms of bank-firm relationships. Though relationships of all forms have effects on commercial bank loan contracting, they certainly do not always lead into the same direction. Bank affiliations formed by dispatched directors offer advantages that seem to secure to some degree the monopoly power for the bank; thus these banks can continue to extract higher-than-competitive interest rates on borrowings. The remaining three kinds of bank-firm relationships we examined, however, have submitted to competition. The analysis of the stability of interest rates shows an equally heterogeneous picture. While main bank affiliations remain able to protect the client firm from fluctuations of interest rates, the remaining three types of bank-firm relationships seem unable to do so.

Comparing our results with previous findings, the analysis further supports the view that the degree of competition in credit markets is an important determinant of the value and effect of bank-firm relationships. As has been already suggested by others (e.g. Kobayashi et al. 1993), the liberalization process seems to make banking relationships more businesslike and therefore more clearly driven by market factors like profitability. Since risk is not significant when included in either of these regressions and overall firm demand for liquidity is declining over the period observed, the evidence most likely suggests a decline in bank monopoly while liberalization proceeds, rather than a declining implicit bankruptcy insurance premium. While a relationship reduces the lender's expected cost, the lender only shares this cost reduction with the firm when competition encroaches upon the lender's informational monopoly. We have shown that the position of banks remains unchanged in areas where banks, for one reason or the other, are not confronted with competition or can cope with competition due to advantages in other areas. Furthermore, our findings are consistent with the financial intermediation literature cited above (e.g. Rajan 1992), which shows that, to mitigate the problem of ex post rent extraction by informed lenders, the borrowing firm will attempt to weaken the bargaining power of these lenders by giving higher priority to arm's-length finance. Obviously, liberalization in Japan increases the firm's capacity to do so. The results suggest that with competition, banks increasingly have to share the cost savings they may accrue from relationships with firms. Though it is obviously increasing, note that we are not able to detect to what degree banks pass these cost savings on to the client firm.

Liberalization may have many other advantages. However, credit market competition is likely to impose constraints on the ability of the firm and creditor to intertemporally share surplus. As the market power of banks decreases, they cannot extract a larger share of the future surplus generated by the firm, as was possible in a regulated and concentrated market. Thus, competition prohibits banks more and more from generating an implicit equity stake in the firm that could enable it to set a lower interest rate for entrepreneurial projects. This finding agrees with that of Mayer (1988) and Rajan (1992), that increased competition in financial markets reduces the value of relationships because it prevents a bank from reaping the rewards of helping the firm at an early stage. It is therefore possible that banks are less and less able to prevent projects with low short-term returns and/or high risk but long-term viability from being prematurely discontinued. Note that this development is especially dangerous to an economy whose financial markets do not sufficiently compensate for it by initiating alternative structures like venture capital markets.

The findings are furthermore supported by much reported evidence that the banking sector's ability to act as a lender of last resort is deteriorating. Clearly, with only a few exceptions, the developments described above make lending relationships less valuable to a firm since it is less likely to get help when most needed. Finally, it is important to note that the already-weakened Japanese banking sector does not only have to deal with increasing competition from the capital markets. To aggravate their situation, banks have to cope with the unique phenomenon that since the burst of the bubble, interest rates are strongly declining while demand is continuously weakening at the same time. This development dramatically shifts the bargaining power in favor of firms, allowing them to play out banks against alternative sources of finance and most recently also against each other.

5.7 Summary and Concluding Remarks

Our analysis highlights central aspects of the role of Japanese bank-firm relationships in times of liberalization and deregulation. Comparing our results with earlier findings, it becomes evident that financial liberalization and deregulation have triggered dramatic changes concerning the position of banks in the financial market and their relationships with large manufacturing firms.

Since liberalization and deregulation have opened up domestic as well as foreign capital markets, easy access to capital is not limited to bank-affiliated firms. While in earlier years firms with close banking ties may have been more profitable or used their advantages to enhance their growth or the compensation of employees, now whether such a relationship benefits a firm or acts as an obstacle depends largely upon the nature of that relationship. Hence, there are two general points made in this chapter. First, in order to assess the economic role of bank affiliations, it is of eminent importance to differentiate between the various ways of establishing and maintaining bank-firm relationships. Second, the rapid development of the Japanese financial market over the last two decades has reduced the monopoly power of banks in a wide range of areas, which has caused dramatic changes concerning the benefits and costs of bank-firm relationships.

Using a variety of proxies which enable us to differentiate between different aspects of bank-firm relationships, the analysis clearly shows that the effect of a bank affiliation can differ substantially. Concerning the opportunity to enhance profitability through advantages of bank affiliations, only when a bank affiliation allows a firm to use large numbers of *keiretsu* loans may the firm be more profitable than its competitors. If at all, bank-firm relationships seem to have a negative effect on the promotion of growth. Moreover, only the advantages that accompany a main bank affiliation seem to improve the compensation of employees. While bank-firm relationships were widely assumed to decrease firm risk, our empirical results do not find any evidence that this can be explained by a main bank affiliation or a strong dependence on *keiretsu* loans. However, a strong influence from banks via bank dispatched directors or large shareholdings seems to have a positive impact on firm risk.

Perhaps the most interesting finding of our study is the impact of credit market liberalization on the cost and stability of interest rates for bank-affiliated firms. Caves and Uekusa (1976) already argued more than twenty years ago that the rationed Japanese capital market, which protects established banks from competition, is the main reason why Japanese banks were able to easily extract rents generated by bank-firm relationships. They demonstrate that in times of limited competition, banks could obtain these rents via high interest rates and via pressure on client firms to use bank loans as the major source for capital inputs. Comparing our results with previous findings, it seems plausible to argue that with increasing liberalization and decreasing demand for finance, it becomes more and more difficult for banks to extract these rents from their client firms. Only in cases where banks can combine the provision of capital with additional services like managerial advice that cannot be offered by alternative financial sources, may banks continue to charge higher-than-competitive interest rates. Overall, however, competition seems to force banks to increasingly share the rents that may originate from advantages of bank-firm relationships by charging competitive interest rates. This development also affects the impact of bank affiliations on the stability of interest rates: as the debt market becomes increasingly competitive, it seems to be more and more difficult to spread the cost of reducing the volatility of interest rates by asking lower-than-competitive interest rates during one period and higher-than-competitive interest rate during another period. Only banks that can spread these costs over multiple products may still be able to insure client firms against interest rate fluctuations.

Clearly, the interpretation of our empirical results is a far cry from the popular views of the bank-dominated Japanese financial system as an engine of growth for highly efficient firms. Taking a closer look at the system of Japanese bank-firm relationships in transition reveals a more complicated scenario, which includes both advantages and at the same time impediments. Obviously, when alternatives to bank finance increase, the negative components of the system of bank-firm relationships become more evident. The results largely accord with

Mayer (1988) and Rajan (1992), who show that increasing competition in financial markets reduces the value of bank-firm relationships. Nevertheless, the results rule out the notion that relationships are of no value, or that relationships and competition are incompatible. Bank financing often involves a long-term relationship that still may help to attenuate problems of asymmetric information that cannot be solved by arm's-length finance. Also note that despite an increase in arm's-length finance, necessary channels like markets for corporate control, which are typical for engineering intervention in a financial system where arm's-length finance is well developed, have just started to evolve in Japan. Especially until these forms of intervention are fully developed, the Japanese economy will continue to rely heavily on control-oriented finance and therefore bank finance, to ensure that firms behave diligently. In other words, the various forms of bank-firm relationships do not necessarily lose their *raison d'être*. Despite competition, bank-firm relationships can still be valuable for both sides. However, the Japanese financial system and with it, bank-firm relationships have entered a new phase of economic development. Banks slowly begin to realize their fading advantages as well as new opportunities. Hence we can conclude that bank affiliations will continue to be an important factor for the international competitiveness of modern Japanese firms, though not necessarily in the same way as was common during the pre-liberalization period. It is therefore necessary to adjust the image of the banking system as it has existed in much of the post-war era in Japan.

Chapter 6

Conclusion

Pressure from the international as well as domestic level to undertake liberalization, which was further urged by the strong fluctuations during the recent economic boom-bust cycle, has insured that the Japanese economy is progressing towards maturity. In spite of many features that are clearly unique to the specific structure of Japan's political, governmental and economic system, it is after all a capitalist society deeply interdependent with other societies of its kind.

During much of the postwar period, Japan's economy impressed the world with its strong economic growth and success. Within an exceptionally short period of time, Japan became one of the highest-ranking industrialized economies. It managed to do so largely through its strong economic ties with individual nations throughout the world. Having established these ties also means, however, that Japan has to increasingly conform itself to the principles of international trade as well as the international movement of capital, and so the Japanese economy has entered its restructuring phase.

Focusing on bank-firm relationships, the key elements of this structural change which have been of interest for the preceding analysis concern the interaction between financial markets, corporate governance, corporate finance and corporate success. Conventional wisdom during most of the postwar period maintained that the close and stable relationships between large Japanese banks and non-financial corporations were at the heart of industrial finance in Japan and insured a wide range of benefits for Japanese firms that led to their competitive advantage. However, the preceding analysis shows that these bank-firm relationships have come under severe strain since the 1980s, as the Japanese financial system experiences wide-ranging financial deregulation and has to cope with an economic boom-bust cycle that triggered a dramatic banking crisis. The advantages of new financing opportunities other than bank debt that have been introduced since the early 1980s, have ultimately abolished most of the monopoly power banks used to be endowed with during times of underdeveloped Japanese capital markets. The reaction of many firms to move away from monitored bank finance toward arm's-length financial sources, suggest that regulation was important in shaping the bank-led financial system in Japan. However, regulation was not the only reason why the bank-centered financial system developed. Despite the fact that liberalization allows an increasing number of firms to issue bonds, many firms still choose to depend, at least to some extend, on bank borrowings and continue to have close relations with banks via various channels that go much beyond the provision of capital. Obviously, those firms still seem to value the benefits of bank ties over the costs that are also implied in maintaining bank-firm relationships and there is no sign that this will change in the near future. However, these ties are not necessarily similar to those prior to the 1980s. To survive in an increasingly unfriendly banking climate, banks were pressured to adjust to this new environment, which has many drawbacks but also offers some opportunities for them to refine their position in Japan's financial markets.

We have used detailed firm-level data on 92 firms of the manufacturing sector for the period between 1985-1998 in order to document how Japanese bank-firm relationships evolved during this period. Providing both theory and empirical evidence, the preceding analysis contributes to further understanding the system of affiliations between banks and firms in a dynamic environment of fundamental restructuring.

Whatever the eventual outcome for the Japanese financial system and bank-firm relationships in particular, one thing should be clear at this point: Japanese bank-firm relationships were never immune from problems, and indeed were ultimately one of the main targets of domestic as well as international pressures on the Japanese financial system. The system of bank-firm relationships is not static, nor does it impart any unfair advantages in the competitive arena of corporate finance. The process of disintermediation has clearly left its marks, and as liberalization proceeds it becomes increasingly clear that close bank-affiliations have not only advantages but also disadvantages for firms. Nevertheless, in the current state of Japanese financial markets, bank-firm relationships remain an important element of Japan's economic system.

Appendix

I Definition of Variables

Variable	Explanation
ACCESS	measure indicating accessibility and usage of bond financing
bank borrowings/total assets	ratio of long and short term borrowings from banks to total assets
bank borrowings/ total debt	ratio of long and short term bank borrowings to total debt
bank borrowings/total liabilities	fraction of bank-financed debt
bonds/total assets	ratio of bond financing to total assets
bond interest/ bond liabilities	ratio of bond interest expenses to bond liabilities
capital/employees	total capital to number of employees
capital stock	capital stock and paid in new capital stock (in billion Yen)
cash ratio	ratio of cash and time deposits to current liabilities
debt/equity	*total debt to total equity
debt/sales	total debt to sales
debt/total assets	leverage

Note: * indicates that this variable has been truncated at the upper and lower percentile in order to avoid outliers.

I Definition of Variables (continued)

Variable	Explanation
DISPATCH	index expressing degree of personal relationship between banks and client firms
employment growth	employment growth of each year (yearly difference of log(employees))
equity/total assets	equity ratio
firm risk	firm risk measured by variance of return on sales over five years at a time
GDP	growth (in per cent) of gross domestic product in real terms
interest	interest rate on borrowings
interest on borrowing/sales	ratio of interest on total borrowings to total sales
interest ratio	interest and discount expenses to liabilities
investment/employees	ratio of investment in plant and equipment to number of employees
investment/sales revenue	ratio of investment in plant and equipment to sales revenue
KEILO	measure indicating existence and degree of *kinyû keiretsu* affiliation under the aspect of loan-dependency

Note: * indicates that this variable has been truncated at the upper and lower percentile in order to avoid outliers.

I Definition of Variables (continued)

Variable	Explanation
KEIRETSU	dummy variable indicating whether or not a firm has a *kinyû keiretsu* affiliation
leverage ratio (mv)	ratio total liabilities to market value
ln(assets)	natural logarithm of firm size measured by total assets
ln(employees)	natural logarithm of firm size measured by number of employees
log sales	logarithm of sales
long-term liabilities/ total assets	ratio of long-term liabilities (maturities longer than one year) to total assets
MB	dummy variable indicating whether or not a firm has a main-bank affiliation
MV	market value of the firm to total liabilities
net income	total net income (in billion Yen)
net income/sales	total net income to total sales
nonbond interest/ nonbond liabilities	ratio of nonbond interest expenses to nonbond liabilities
ordinary income/ sales	ordinary income to sales
profit/interest	*ratio of pre-tax profit (loss) to total interest payments
profit/total assets	pretax profits to total assets

Note: * indicates that this variable has been truncated at the upper and lower percentile in order to avoid outliers.

I Definition of Variables (continued)

Variable	Explanation
propensity to invest	ratio of investment in fixed assets to fixed assets of previous year
RME	ratio of male employees to total employees adjusted with respect to age structure of firm's work force: $$RME = \frac{ME \times AGM}{ME \times AGM + FE \times AGF}$$ with ME: number of male employees; AGM: average age of male employees; FE: number of female employees; AGF: average age of female employees
ROE	*after-tax return on equity
ROI	return on investment
ROS	return on sales
sales growth	annual growth rate of sales
SHARE	ratio of client-firm-shares held by banks among the 20 largest shareholders
short-term liabilities/total assets	ratio of short-term liabilities (maturities up to one year) to total assets
size	variable ranging from 1 (smallest) to 4 (largest) defined by quartiles of total assets
tangible fixed asset growth	*growth rate of tangible fixed assets

Note: * indicates that this variable has been truncated at the upper and lower percentile in order to avoid outliers.

I Definition of Variables (continued)

Variable	Explanation
total labor cost/employees	firm's total labor cost (including salary, other compensations, contributions to fringe benefits and social security) to the number of employees at the end of each accounting period
var interest rate; prime rate	variance of difference between interest rate and prime rate over five years increments

Note: * indicates that this variable has been truncated at the upper and lower percentile in order to avoid outliers.

II Descriptive Statistics for Regression Variables

Variable	N	Mean	25 % Percentile	75% Percentile
ACCESS	1288	3.897	2.423	5.363
bank borrowing/ total assets	1288	0.139	0.018	0.217
bank borrowings/total debt	1288	0.444	0.121	0.735
bank borrowings/ total liabilities	1288	0.203	0.040	0.328
bonds/total assets	1288	0.121	0.053	0.180
bond interest/ bond liabilities	1007	0.031	0.017	0.039
capital/employees	1288	66.234	43.841	82.211
capital stock	1288	43.300	10.800	43.200
cash ratio	1288	0.460	0.158	0.516
debt/equity	1288	2.006	0.854	2.553
debt/sales	1288	0.732	0.549	0.870
debt/total assets	1288	0.582	0.461	0.719
DISPATCH	1288	4.744	0.000	7.326
employment growth	1288	0.001	-0.022	0.026
equity/total assets	1288	0.418	0.281	0.539
firm risk	1288	0.021	0.009	0.024
GDP	1288	2.750	1.000	4.800

Source: KAISHA database; author's computation

II Descriptive Statistics for Regression Variables (continued)

Variable	N	Mean	25 % Percentile	75% Percentile
interest	1288	0.025	0.016	0.032
interest on borrowings/ sales	1288	0.019	0.010	0.024
interest ratio	1288	0.016	0.006	0.023
investment/employees	1288	5326.641	2240.813	6800.329
investment/sales revenue	1288	0.102	0.054	0.131
KEILO	1288	0.115	0.000	0.239
KEIRETSU	1288	0.402	---	---
leverage ratio (mv)	1288	0.384	0.263	0.503
ln(assets)	1288	19.317	18.542	20.049
ln(employees)	1288	8.320	7.574	8.845
log sales	1288	19.106	18.510	19.819
long-term liabilities/ total assets	1288	0.190	0.128	0.254
MB	1288	0.814	---	---
MV	1288	3.340	1.988	3.807
net income	1288	10.100	1.600	10.800
net income/sales	1288	0.024	0.012	0.038

Source: KAISHA database; author's computation

II Descriptive Statistics for Regression Variables (continued)

Variable	N	Mean	25 % Percentile	75% Percentile
nonbond interest/ nonbond liabilities	1271	0.020	0.009	0.029
ordinary income/sales	1288	0.024	0.012	0.038
profit/interest	1288	7.041	1.001	6.967
profit/total assets	1288	0.043	0.018	0.067
propensity to invest	1288	0.418	0.229	0.540
RME	1288	0.884	0.848	0.929
ROE	1288	0.046	0.029	0.081
ROI	1288	0.035	0.025	0.049
ROS	1288	0.056	0.025	0.079
sales growth	1288	0.038	-0.020	0.092
SHARE	1288	0.218	0.162	0.268
short-term liabilities/ total assets	1288	0.391	0.276	0.498
size	1288	2.498	---	---
tangible fixed asset growth	1288	0.065	-0.016	0.113
total labour cost/ employees	1288	6.396	5.585	7.387
var interest rate; prime rate	1288	1.019	0.572	1.408

Source: KAISHA database; author's computation

III Development of Balance-Sheet Items: Full Sample, 92 Firms[75]

	FY1985	FY1989	FY1993	FY1997
Assets	100.00	100.00	100.00	100.00
Current assets	67.05	64.97	57.80	58.55
Cash and deposits	12.16	16.86	11.79	9.15
Trade receivables	28.91	25.90	23.30	24.85
Short-term securities	6.09	6.36	6.72	8.55
Products and raw materials	16.53	12.87	12.59	12.34
Short term loans	0.21	0.19	0.22	0.39
Other	3.15	2.79	3.18	3.27
Fixed assets	32.95	35.03	42.20	41.45
Tangible fixed assets	20.51	21.34	24.84	23.35
Of which Depreciable tangible fixed assets	15.76	15.85	18.75	17.04
Land	3.51	4.29	4.81	5.14
Other	1.24	1.20	1.28	1.17
Intangible fixed assets	0.20	0.23	0.23	0.21
Investments and other assets	12.23	13.46	17.13	17.89
Of which Investment securities	7.95	9.61	12.84	14.38
Loans receivable	2.66	2.02	2.09	1.48
Other	1.62	1.83	2.20	2.03
Deferred Assets	0.01	0.00	0.00	0.00
Liabilities	64.09	59.09	56.71	55.54
Current liabilities	48.04	40.76	36.11	37.00
Of which Notes payable	21.34	16.98	13.86	15.38
Short-term loans payable	11.99	12.58	13.21	8.80
Accrued taxes	2.92	2.69	1.13	1.40
Current provisions	0.14	0.12	0.13	0.12
Other	11.65	8.39	7.78	11.30
Fixed liabilities	16.05	18.33	20.60	18.54
Of which Long-term loans payable	5.69	3.34	4.80	3.95
Bonds payable	6.32	10.88	12.05	10.79
Fixed provisions	3.61	3.43	3.41	3.28
Other	0.43	0.68	0.34	0.52
Equity capital	35.91	40.91	43.29	44.46
Capital stock	7.60	9.78	10.58	10.94
Capital reserves	9.42	11.30	11.66	11.90
Legal reserves	1.12	1.15	1.30	1.44
Other retained earnings	17.77	18.68	19.75	20.18

[75] As a percentage of total assets; Source: KAISHA database; authors' computation

IV Sample Statistics of Bank-Affiliation Criteria[76]

1985 – 1998

VARIABLE NAME	OBS.	MIN	MEAN	MEDIAN	MAX	STD.DEV.
DISPATCH	1288	0.00	4.74	2.17	41.94	6.57
MB	1288	0.00	0.81	1.00	1.00	0.39
KEILO	1288	0.00	7.90	0.00	45.25	12.16
SHARE	1288	5.42	21.85	21.49	45.33	7.48

1985 - 1988

VARIABLE NAME	OBS.	MIN	MEAN	MEDIAN	MAX	STD.DEV.
DISPATCH	368	0.00	5.41	2.74	41.94	7.67
MB	368	0.00	0.83	1.00	1.00	0.38
KEILO	368	0.00	7.76	0.00	43.63	12.03
SHARE	368	5.42	17.59	16.17	32.06	6.36

1989 – 1992

VARIABLE NAME	OBS.	MIN	MEAN	MEDIAN	MAX	STD.DEV.
DISPATCH	368	0.00	4.85	2.65	31.43	6.01
MB	368	0.00	0.85	1.00	1.00	0.36
KEILO	368	0.00	7.96	0.00	38.61	12.21
SHARE	368	6.61	22.96	22.81	44.17	7.95

1993 – 1996

VARIABLE NAME	OBS.	MIN	MEAN	MEDIAN	MAX	STD.DEV.
DISPATCH	368	0.00	4.16	1.65	34.38	6.05
MB	368	0.00	0.77	1.00	1.00	0.42
KEILO	368	0.00	7.96	0.00	45.25	12.40
SHARE	368	12.31	24.18	24.72	45.33	6.35

1997 – 1998

VARIABLE NAME	OBS.	MIN	MEAN	MEDIAN	MAX	STD.DEV.
DISPATCH	184	0.00	4.36	1.86	25.68	6.16
MB	184	0.00	0.80	1.00	1.00	0.40
KEILO	184	0.00	7.92	0.00	41.10	12.28
SHARE	184	6.70	23.48	22.53	44.16	7.19

[76] Source: KAISHA database; authors' computation

V OLS-Regression Results

V.a Evidence of Bank-Affiliation for Amount of Bank-Financed Debt

PARAMETER	COEFFICIENT (T-TEST)
DISPATCH	0.00618*** (9.027)
SHARE	-0.00575*** (-8.927)
Industry dummies	Yes
Year dummies	Yes
N	1288
Adj. R^2	0.2845

Source: KAISHA database, authors' calculation
*** significant at a level ≤ 1 %
The dependent variable is the amount of bank debt normalized by total liabilities, and the regression is OLS using pooled data from the entire sample.

V.b Evidence of Bank-Affiliation for Firm's Market Value

PARAMETER	COEFFICIENT (T-TEST)
DISPATCH	-0.02548*** (-2.920)
SHARE	0.06211*** (7.561)
Industry dummies	Yes
Year dummies	Yes
N	1288
Adj. R^2	0.2392

Source: KAISHA database, authors' calculation
*** significant at a level ≤ 1 %
The dependent variable is the firm's market value normalized by total liabilities, and the regression is OLS using pooled data from the entire sample.

VI List of Firms

Nr.	Security Code	Company	Kinyû keiretsu	Industry
1	6118	Aida Engineering	Non	Machinery
2	6113	Amada	Non	Machinery
3	6107	Amada Sonoike	Non	Machinery
4	6108	Amada Wasino	Non	Machinery
5	6140	Asahi Diamond Industrial	Non	Machinery
6	4044	Central Glass	Non	Chemicals
7	6366	Chiyoda Corporation	Non	Machinery
8	4519	Chugai Pharmaceutical	Non	Pharmaceuticals
9	6796	Clarion	Non	Electric Machinery
10	4202	Daicel Chemical Industries	Non	Chemicals
11	6383	Daifuku	Non	Machinery
12	4505	Daiichi Pharmaceutical	Non	Pharmaceuticals
13	6367	Daikin Industries	Non	Machinery
14	4116	Dainichiseika Color & Chemicals Mfg.	Non	Chemicals
15	4061	Denki Kagaku Kogyo	Daiichi	Chemicals
16	6138	Dijet Industrial	Non	Machinery
17	6361	Ebara	Daiichi	Machinery
18	4523	Eisai	Non	Pharmaceuticals
19	6504	Fuji Electric	Daiichi	Electric Machinery
20	4511	Fujisawa Pharmaceutical	Sanwa	Pharmaceuticals
21	6702	Fujitsu	Daiichi	Electric Machinery
22	6755	Fujitsu General	Non	Electric Machinery
23	6501	Hitachi	Daiichi	Electric Machinery
24	6581	Hitachi Koki	Non	Electric Machinery
25	6106	Hitachi Seiki	Non	Machinery
26	6102	Ikegai	Non	Machinery

VI List of Firms (continued)

Nr.	Security Code	Company	Kinyû keiretsu	Industry
27	6310	Iseki &Co.	Daiichi	Machinery
28	7013	Ishikawajima-Harima Heavy Industries	Daiichi	Transport Equipment
29	6751	Japan Radio	Non	Electric Machinery
30	4185	Japan Synthetic Rubber	Non	Chemicals
31	4118	Kaneka	Non	Chemicals
32	4452	Kao	Non	Chemicals
33	7012	Kawasaki Heavy Industries	Daiichi	Transport Equipment
34	6765	Kenwood	Non	Electric Machinery
35	6756	Kokusai Electric	Non	Electric Machinery
36	6301	Komatsu	Non	Machinery
37	6326	Kubota	Fuyô	Machinery
38	4151	Kyowa Hakko Kogyo	Daiichi*	Chemicals
39	6135	Makino Milling Machine	Non	Machinery
40	6586	Makita	Non	Electric Machinery
41	6752	Matsushita Electric Industrial	Non	Electric Machinery
42	6508	Meidensha	Non	Electric Machinery
43	6503	Mitsubishi Electric	Mitsubishi	Electric Machinery
44	4182	Mitsubishi Gas Chemical	Mitsubishi	Chemicals
45	7011	Mitsubishi Heavy Industries	Mitsubishi	Transport Equipment
46	4010	Mitsubishi Kasei	Mitsubishi	Chemicals
56	6103	Okuma	Non	Machinery
57	6645	Omron	Non	Electric Machinery
58	6644	Osaki Electric	Non	Electric Machinery

*Kyowa Hakko Kogyo Co. (4151) is member of the Daiichi group since 1988.

VI List of Firms (continued)

Nr.	Security Code	Company	Kinyû keiretsu	Industry
59	6136	OSG Mfg.	Non	Machinery
60	6773	Pioneer Electronic	Non	Electric Machinery
61	6707	Sanken Electric	Non	Electric Machinery
62	4501	Sankyo	Daiichi	Pharmaceuticals
63	6764	Sanyo Electric	Non	Electric Machinery
64	4204	Sekisui Chemical	Sanwa	Chemicals
65	6753	Sharp	Sanwa	Electric Machinery
66	4063	Shin-Etsu Chemical	Non	Chemicals
67	4507	Shionogi	Non	Pharmaceuticals
68	4004	Showa Denko	Fuyô	Chemicals
69	6758	Sony	Non	Electric Machinery
70	4203	Sumitomo Bakelite	Sumitomo	Chemicals
71	4005	Sumitomo Chemical	Sumitomo	Chemicals
72	6302	Sumitomo Heavy Industries	Sumitomo	Machinery
73	6395	Tadano	Non	Machinery
74	4535	Taisho Pharmaceutical	Non	Pharmaceuticals
75	4502	Takeda Chemical Industries	Non	Pharmaceuticals
76	4508	Tanabe Seiyaku	Sanwa	Pharmaceuticals
77	6803	Teac	Non	Electric Machinery
78	4045	Toagosei Chemical Industry	Non	Chemicals
79	4043	Tokuyama Soda	Sanwa	Chemicals
80	6502	Toshiba	Mitsui	Electric Machinery
81	6104	Toshiba Machine	Non	Machinery
82	6139	Toshiba Tungaloy	Non	Machinery
83	4042	Tosoh	Non	Chemicals

VI List of Firms (continued)

Nr.	Security Code	Company	Kinyû keiretsu	Industry
84	6330	Toyo Engineering	Non	Machinery
85	6201	Toyoda Automatic Loom Works	Non	Machinery
86	6206	Toyoda Machine Works	Non	Machinery
87	6101	Tsugami	Non	Machinery
88	4540	Tsumura & Co	Non	Pharmaceuticals
89	4225	Tsutsunaka Plastic Industry	Non	Chemicals
90	4208	Ube Industries	Sanwa	Chemicals
91	4503	Yamanouchi Pharmaceutical	Non	Pharmaceuticals
92	6506	Yasukawa Electric	Daiichi	Electric Machinery

Abbreviations

BIS	Bank of International Settlement
BOJ	Bank of Japan
CD	Certificates of Deposits
DAAD	Deutscher Akademischer Austauschdienst
DFG	Deutsche Forschungsgemeinschaft
DKB	Dai Ichi Kangyô Bank
dpa	Deutsche Presseargentur
EEA	European Economic Association
FEFTCL	Foreign Exchange and Foreign Control Law
GDP	Gross Domestic Product
IBJ	Industrial Bank of Japan
KKS	Kigyô Keiretsu Soran (source of reference for *keiretsu* firms)
MMC	Money Market Certificates
MOF	Ministry of Finance
Nikkei	Nihon Keizai Shinbun (Newspaper)
NTT	Nippon Telegraph and Telephone
UK	United Kingdom
WHU	Wissenschaftliche Hochschule für Unternehmensführung
WZB	Berlin Science Center (Wissenschaftszentrum Berlin)

Figures and Tables

Bibliography

Abegglen, J. and Stalk, G. (1985). *Kaisha, The Japanese Corporation*, New York, NY, Basic Books, Inc.

Abrahams, P. and Hess, J. (1998). Alarm bells ring over total group debt, *Financial Times*, 28.10.1998, page 16.

Albach, H. (1999a). *Flexibility and Growth – An Analysis of Crisis Managing in Japanese Corporations*, Kyôtô, International Institute for Advanced Studies.

Albach, H. (1999b). *Corporate Governance and Bank-Firm Relationship*, in: The Finnish School of Business: The Jaakko-Honko-Lectures, Helsinki.

Albach, H. (1998). Bankenbeziehungen und Wettbewerbsfähigkeit, *ZfB*, Ergänzungsheft 2/98, 1-27.

Albach, H., Brandt, S., Brandt, T., and Yang, J. (1996). *Bank-Firm Relationships and Corporate Performance.* Heidelberg, Paper submitted to the Conference on Input-Related Industrial Economics.

Albach, H., Görtzen, U., Miarka, T., Moerke, A., Westpahl, T. and Zobel, R. (1997). *Dokumentation der Kaisha-Datenbank – Zur Datenbank der Jahresabschlüsse japanischer Aktiengesellschaften.* Berlin, WZB Discussion paper FS IV 97-39, Berlin Science Center.

Albach, H. and Miarka, T. (1997). *Case: Nomura Securities*, prepared for the Kellogg-WHU Executive MBA session on "Governance Structures Compared".

Amemiya, T. (1971). The estimation of the variance in a variance-components model. *International Economic Review.* 12, 1-13.

Anderson, C. W. and Makhija, A. K. (1999). Deregulation, disintermediation, and agency costs of debt: evidence from Japan, *Journal of Financial Economics*, 51, 309-339.

Aoki, M. (1987). The Japanese firm in Transition. In K. Yamamura and Y. Yasuba (Eds.), *The Political Economy of Japan.* Stanford, CA: Stanford University Press, 263-288.

Aoki, M. (1998): National Diversity of Corporate Governance and the Global Standard, transcript of the opening remarks by Masahiko Aoki at the Fifth International Conference Europe-Japan held in Spain, December 1997, in: *Journal of Japanese Trade & Industry*, September/October, 20-23.

Aoki, M. and Patrick, H. (Eds.) (1994). *The Japanese Main Bank System*. Oxford, Oxford University Press.

Aoki, M., Patrick, H. and Sheard, P. (1994). The Japanese main bank system: An introductory overview. In M. Aoki and H. Patrick (Eds.), *The Japanese Main Bank System*, Oxford, Oxford University Press.

Association of Securities Underwriters. *Monthly Report of Japanese Bonds*, Tôkyô, various issues.

Ballon, R. and Tomia, I. (1988). *The Financial Behavior of Japanese Corporations*, New York, NY, Kodansha International.

Baltagi, B. (1995). *Econometric Analysis of Panel Data*. Chichester, John Wiley & Sons.

Baltagi, B. and Li, Q. (1990). A Lagrange multiplier test for the error components model with incomplete panels. *Econometric Reviews*, 9 (1), 103-107.

Bank of Japan (1992). *Analysis of Recent Changes in the Relationship between Banks and Corporations Based on Financial Data of Corporations*, Special paper No. 217, July 1992.

Bank of Japan (1996). Changes in the Corporate Balance-Sheet Structure since 1980 in Japan, In *Bank of Japan Quarterly Bulletin*, November 1996, 59-98.

Bank of Japan. *Shyô kigyô keiei bunseki* [Analysis of Financial Statements of Principle Enterprises], several issues.

Barnea, A., Haugen R., and Senbet, L. (1985). *Agency Problems and Financial Contracting*. Englewood Cliffs, NJ, Prentice-Hall.

Baron, P. P. (2000). *Big Bang from the Foreign Perspectives*, in Albach et. al. (eds.): "Evaluation of Firms, Performance and Financial Structure" (forthcoming), Tôkyô, Waseda University.

Beason, D. and James, J. (1999). *The political economy of Japanese financial markets: myths versus reality*. New York, NY, St. Martin's Press.

Ben-Porath, Y. (1973). Labor Force Participation Rates and the Supply of Labor. *Journal of Political Economy* 81, 697-704.

Berger, A. N. and Udell, G. (1995). Relationship Lending and Lines of Credit in Small Firm Finance, *Journal of Business*, Vol. 68, 351-381.

Berglöf, E. and Sjörgen, H. (1998). Combining Arm's-Length and Control-Oriented Finance - Evidence from Main Bank Relationships in Sweden, in Hopt, K.; Kanda, H.; Roe, Wymeersch, M.; and Prigge, S. (eds.), *Comparative Corporate Governance – The State of the Art and Emerging Research*, Oxford, Oxford University Press, 787-808.

Berlin M. and Loeys, J. (1988). Bond covenants and delegated monitoring, *Journal of Finance*, 43, 397-412.

Bhargava, A.; Franzini, L. and Narendranathan W. (1982). Serial Correlation and the Fixed Effects Model, in *Review of Economic Studies,* XLIX, 533-549.

Billett, M. T.; Flannery, M. J. and Garfinkel, J. A. (1995). The effect of lender identity on a borrowing firm's equity return, *Journal of Finance*, 50, 699-718.

Black, B. S. (1992). Agents Watching Agents: The Promise of Institutional Investor Voice, *UCLA Law Review*, 39, 811-893.

Blackwell, N. and Santomero, A. (1982). Bank credit rationing and customer relation, *Journal of Monetary Economics*, 9, 121-129.

Borio, C., Kennedy, N., and Prowse, S. (1994). Exploring Aggregate Asset Price Fluctuation across Countries, *BIS Economic Papers No. 40.*

Boyd, J. and Prescott, E. C. (1986). Financial intermediary-coalitions, *Journal of Economic Theory*, 38, 211-232.

Brainard, W. and Tobin, J. (1963). Financial Intermediaries and the Effectiveness of Monetary Control, *American Economic Review*, 53, 383-400.

Breusch, T. and Pagan, A. (1980) The Lagrange multiplier test and its applications to model specification in ecbnomitrics, *Review of Economic Studies*, 47, 239-253.

Campbell, T. and Kracaw, W. (1980). Information production, market signaling, and the theory of intermediation, *Journal of Finance*, 35, 863-882.

Cargill, T., Hutchinson, M., and Ito, T.(1997). Japan's "Big Bang" Financial Deregulation: Implications for Regulatory and Supervisory Policy, *Japan Information Access Project Proceedings on Japanese Deregulation,* http://www.nmjc.org/jiap/frame.html.

Caves, R. and Uekusa, M. (1976). *Industrial Organization in Japan.* The Brookings Institution, Washington, D.C.

Chemmanur, T. and Fulghieri, P. (1994). Reputation, renegotiation, and the choice between bank loans and publicly traded debt, *Review of Financial Studies*, 7, 475-506.

Chirinko, R. (1997). Finance Constraints, Liquidity, and Investment Spending: Theoretical Restrictions and International Evidence, *Journal of the Japanese and International Economies*, 11, 185-207.

Chow, G. (1960). Tests of Equality Between Sets of Coefficients in Two Linear Regressions. *Econometrica*, 28, 591-605.

Choy, J. (1994). *Update on Deregulation of Japan's Financial Markets*, Japan Economic Institute Report, May 20.

Choy, J. (1998). Japan's Financial Market Big Bang: The First Shock Waves, Japan Economic Institute Report, No. 22A, June12.

Coenenberg, A.G. (1991). *Jahresabschluß und Jahresabschlußanalyse*, 13. Edition, Landsberg am Lech.

Corbett, J. (1987). International perspective on financing: evidence from Japan, *Oxford Review of Economic Policy*, 3, 30-55.

Degryse, H. and Ongena, S. (1999). Bank Relationships and Firm Profitability, paper presented at EEA99.

Diamond, D. W. (1984). Financial intermediation and delegated monitoring, *Review of Economic Studies,* 51, 393-414.

Diamond, D. W. (1991). Debt Maturity Structure and Liquidity Risk, *Quarterly Journal of Economics,* Vol. 56, 709-738.

Dolles, H. (1997). *Keiretsu. Emergenz, Struktur, Wettbewerbsstärke und Dynamik japanischer Verbundgruppen.* Europäische Hochschulschriften, Frankfurt a.M., Peter Lang Verlag.

Elston, J., and Albach, H. (1995). Bank Affiliations and Firm Capital Investment in Germany, *ifo Studien*, 41, 3-16.

Fama, E. (1980). Banking in the Theory of Finance, *Journal of Monetary Economics*, 6, 39-57.

Fama, E. (1985). What's different about banks?, *Journal of Monetary Economics*, 15, 29-36.

Farmer, R. (1984). New theory of aggregate supply, *American Economic Review*, 74 (5), 920-930.

Fazzari, S., Hubbart, R., and Peterson, B. (1988). Financing Constraints and Corporate Investment, *Brookings Papers on Economic Activity*, 1, 141-195.

Fisher, F. (1987). On the misuse of the profit-sales ratio to infer monopoly power, *RAND Journal of Economics,* 18, 384-396.

Fisher, I (1933). The debt-deflation theory of great depressions, *Econometrica,*1 (October), 337-357.

Frankel, J. A. (1984). *The yen/dollar agreement: Liberalizing Japanese capital markets,* Policy Analysis in International Economics no. 9, Washington, D.C., Institute for International Economics.

Freixas, X., and Rochet, J.-C. (1997). *Microeconomics of Banking*, Cambridge, MIT Press.

Fukao, M. (1995). Financial Integration, Corporate Governance, and the Performance of Multinational Companies, Washington D.C., The Brookings Institution.

Gale, D., and Hellwig, M. (1985). Incentive-Compatible Debt Contracts: The One-Period Problem, *Review of Economic Studies*, 52, 647-664.

Geisen, B. (1979). *Das Finanzierungsverhalten deutscher Industrieaktiengesellschaften*, in Albach, H.; Krümmel, H.-J.; and Sabel, H. (eds.), Bonner Betriebswirtschaftliche Schriften, Bonn.

Georg, S. (1996). *Die Leistungsfähigkeit japanischer Banken. Eine Strukturanalyse des Bankensystems in Japan.* Berlin Science Center, Unit: Market Dynamics (ed.), Berlin: Edition Sigma.

Gerlach, M. (1992). *Alliance Capitalism. The Social Organization of Japanese Business.* Berkley and Los Angeles, CA: University of California Press.

Gorton, G. and Haubrich, J.(1986). *Bank deregulation, credit markets, and the control of capital,* mimeo, The Wharton School, University of Pennsylvania.

Göseke, C. (1996). Content of Annual Reports of Listed Japanese Corporations, paper given at the Workshop "Rechnungslegung und Analyse japanischer Unternehmen" 10./11.10.1996, Berlin, mimeo.

Greenbaum, S.; Kanatas, G.; and Venezia, I. (1989). Equilibrium loan pricing under the bank client relationship, *Journal of Banking and Finance,* 13, 9-20.

Greene, W. H. (1993): *Econometric Analysis.* Prentice-Hall, Upper Saddle River, NJ.

Greenwald, B., and Stiglitz, J. (1993). Financial Market Imperfections and Business Cycles, *Quarterly Journal of Finance,* 50, 1377-1419.

Gurley, J., and Shaw, E. (1955). Financial aspects of economic development, *American Economic Review,* 65, 515-538.

Hadley, E. M. (1970). *Antitrust in Japan,* Princeton, Princeton University Press.

Hall, M. (1998). *Financial Reform in Japan,* Northampton, Mass., Edward Elgar Publishing.

Harhoff, D., Stahl, K., and Woywode, M. (1998). Legal Form, Growth and Exit of West German Firms. Empirical Results for Manufacturing, Construction, Trade and Service Industries, *Journal of Industrial Economics,* 46(4), 453-488.

Harhoff, D. and Körting, T. (1998a). How Many Creditors Does it Take to Tango?, mimeo, Berlin Science Center.

Harhoff, D. and Körting, T. (1998b). Lending relationships in Germany – Empirical evidence from survey data, *Journal of Banking and Finance,* (22) 10-11, 1317-1353.

Haubrich, J. (1989). Financial intermediation, delegated monitoring, and long-term relationships, Journal of Banking and Finance, 13, 9-20.

Hausman, J. A. (1978). Specification tests in econometrics. *Econometrica,* 46, 1251-1271.

Hirata, M. (1995). Die japanische torishimariyaku-kai: Eine rechtliche und betriebswirtschaftliche Analyse, *Hitotsubashi Journal of Commerce and Management* 30-1, 1-27. (In Germany this article has been published in *Zeitschrift für Betriebswirtschaftslehre (zfb)*, EH 3/96, 1-27).

Hodder, J. and Tschoegl, A. (1985). Some aspects of Japanese corporate finance, *Journal of Finance and Quantitative Analysis*, 20, 173-191.

Hodder, J. and Tschoegl, A. (1993). Corporate financing in Japan, in Tagaki, S. (ed.), *Japanese Capital Markets*, Cambridge, MA, Basil Blackwell.

Holmström, B. and Tirole, J. (1997). Financial Intermediation, Loanable Funds, and the Real Sector. *Quarterly Journal of Economics*, Vol. CXII. Issue 3, 663 – 691.

Horiuchi, A. (1996). An Evaluation of Japanese Financial Liberalization: A Case Study of Corporate Bond Markets, in: Ito, T and Krueger, O (eds.), *Financial deregulation and integration in East Asia*, Chicago, The University of Chicago Press, 167-192.

Hoshi, T. (1993). *Evolution of the main bank system in Japan*, mimeo, University of California, San Diego.

Hoshi, T. (1995). *Financial deregulation and corporate financing in Japan*, mimeo, La Jolla: Graduate School of International Relations and Pacific Studies, University of California.

Hoshi, T.; Kashyap, A.; and Loveman, G. (1994). Lessons from the Japanese main bank system for financial reform in Poland, in Aoki, M. and Patrick, H. (eds.), *The Japanese Main Bank System*, Oxford, Oxford University Press, 592-633.

Hoshi, T.; Kashyap, A. and Scharfstein, D. (1990a). Bank monitoring and investment: Evidence from the changing structure of Japanese corporate banking relationships, in Hubbard (ed.) *Asymmetric Information, Corporate Finance and Investment,* Chicago, University of Chicago Press, 105-126.

Hoshi, T.; Kashyap, A. and Scharfstein, D. (1990b). The role of banks in reducing the costs of financial distress in Japan, *Journal of financial Economics*, 27, 67-88.

Hoshi, T.; Kashyap, A. and Scharfstein, D. (1991). Corporate structure, liquidity and investment: evidence from Japanese industrial groups, *Quarterly Journal of Economics*, 106, 33-60.

Houston, J. and James, C. (1996). Bank information monopolies and the mix of private and public debt claims, *Journal of Finance*, 51, 1863-1889.

Hsiao, C. (1986). *Analysis of panel data*, Cambridge, Cambridge University Press.

Itami, H. (1998). *Nihon sangyô mitsu no tsunami (The three waves of Japanese industry)*, Tôkyô, NTT Press.

Itô, T. and Hoshi, T. (1992). Kigyô gurûpu kessokudo no bunseki (Analysis of circumstantial evidence of interconnections in company groups), in: Horiuchi, A, and Yoshino, N. (eds.) *Gendai nihon no kinyû bunseki*, Tôkyô: Tôkyô Daigaku Shuppankai, 73-96.

Itô, T., and Iwaisako, T. (1996). Explaining Asset Bubbles in Japan. *BOJ Monetary and Economic Studies*, Vol. 14 No. 1, 143-193.

James, C. (1987). Some evidence on the uniqueness of bank loans, *Journal of Financial Economics*, 19, 217-235.

James, C. and Wier, P. (1990). Borrowing relationships, intermediation, and the cost of issuing public securities, *Journal of Financial Economics*, 28, 149-171.

Japan Economic Institute (1993). *The Cooperative Credit Purchasing Co.*, JEI Report No. 4B, February 5, 1993.

Jensen, M. and Meckling, N. (1976). Theory of the firm: Managerial behavior, agency costs and ownership structure. *Journal of Financial Economics*, 3(4), 305-360.

Kang, S. (1985). A note on the equivalence of specification tests in the two-factor multivariate variance components model. *Journal of Econometrics* 28, 193-203.

Kaserer, C. and Wenger, E. (1998). German Banks and Corporate Governance - A Critical View in: Hopt, K.J. et al. (ed.): *Comparative Corporate Governance - The State of the Art and Emerging Research*, Oxford, 499-536.

Kashyap, A., Lamont, O., and Stein, J. (1994). Credit Conditions and the Behaviour of Inventories, *Quarterly Journal of Economics*, 109, 565-592.

Kashyap, A., Stein, J., and Wilcox, D. (1993). Monetary Policy and Credit Conditions: Evidence from the Composition of External Finance, *American Economic Review*, 83, 78-98.

Keizai kikaku chô (1992). Keizai Hakusho (Economic White Paper) 1992, Tokyo: Ôkurashô insatsujo hakkô.

Kennedy, P. (1985). *A Guide to Econometrics*, 2. Edition, Cambridge, Mass., MIT Press.

Kester, W. K. (1991). *Japanese Takeovers: The Global Contest for Corporate Control*, Boston, Mass., Harvard Business School Press.

Kiyotaki, N., and Moore, J. (1997). Credit Cycles, *Journal of Political Economy*,105 No. 21, 213-248.

Kobayashi, H., Endo, Y., and Ogishima, S. (1993). New Directions in Japanese Banking Relationships, Nomura Research Institute, NRI Quarterly-SPRING 1993, 2-27.

Komiya, R. (1990). *The Japanese Economy, Trade, Industry, and Government*, Tôkyô: University of Tôkyô Press.

Kôsei torihiki iinaki (1992). *Nihon no roku dai kigyô shûdan* (The Six Big Japanese Corporate Groups), Tôkyô, Tôkyô keizai shinpôsha.

Kôshasai hikiuke kyôkai (1980). *Nihon kôshasai shijôshi* (History of Public and Corporate Bonds Markets in Japan), Tôkyô, Kôshasai hikiuke kyôkai.

KPMG Peat Marwick (1993). *Banking in Japan*, 8th Edition, Tokyo.

Krishnamurthy, A. (1998). *Collateral Constraints and the Amplification Mechanism*, unpublished paper, Northwestern University.

Kwan, S. H. (1994). *The certification value of bank loans*, Working paper, Tucson, University of Arizona, April.

Lamont, O. (1995). Corporate-Debt Overhang and Macroeconomic Expectations, *American Economic Review*, 85, 1106-1117.

Leamer, E. E. (1978). *Specification Searches: Ad Hoc Inference with Nonexperimental Data*, New York, Wiley.

Leland, H. and Pyle, D. (1977). Information asymmetries, financial structure, and financial intermediaries, *Journal of Finance*, 32, 371-387.

Lincoln, E. J. (1998). Japan's Financial Mess, *Foreign Affairs*, 77 No. 3 (May/June), 57-66.

Lincoln, J. R., Gerlach, M. L., and Takahashi, P. (1992). Keiretsu Networks in the Japanese Economy: A Dyad Analysis of Intercorporate Ties, *American Sociological Review*, Vol. 57, 561-585.

Lummer, S. L. and McConnell, J. J. (1989). Further evidence on the bank lending process and the capital market response to bank loan agreements. *Journal of Financial Economics*, 25, 99-122.

Mankiw, G. (1986). The allocation of credit and financial collapse, *Quarterly Journal of Economics*, 101 (3), 455-470.

Mayer, C. (1990). Financial systems, corporate finance, and economic development, in Hubbard, R. (ed.) *Asymmetric Information, Corporate Finance and Investment*, Chicago, University Press of Chicago.

Meerschwam, D. (1991). *Financial Boundaries: Global Capital, National Deregulation and Financial Service Firms*, Cambridge, MA, Harvard Business School Press.

Miarka, T. (2000). *The Recent Economic Role of Bank-Firm Relationships in Japan*, Berlin, WZB Discussion paper FS IV 99-46, Berlin Science Center.

Miarka, T. (1998). *The notation and procedure for developing measures of bank-firm relationships.* unpublished manuscript, Berlin, Berlin Science Center.

Miarka, T. and Yang, J. (1999). Japanese Bank-Firm-Relationships and Corporate Performance, in Albach, H., Görtzen, U., and Zobel, R. (eds.), *Information Processing as a Competitive Advantage of Japanese Firms*, Berlin, Edition Sigma, 55-79.

Ministry of Finance (Japan). *Ministry of Finance Monthly Finance Review*, several issues.

Ministry of Finance (Japan). *Financial Statements of Incorporated Businesses*, several issues.

Ministry of Finance (Japan), (1997). *Concerning the Amendment of the Foreign Exchange and Foreign Trade Law: Toward the Further Activation of the Financial and Capital Market in Japan*, provisional English translation of the report of the Committee on Foreign Exchange and Other Transactions, Tôkyô, January.

Miwa, Y. (1985). *Mainbank to sono kinô* (The Function of the Main Bank), in: Kosai, Y. and Nishikawa, S. (eds.) Nihon Keizai Sistemu, Tokyo: University of Tokyo Press, 170-199.

Miwa, Y. (1991). Mainbank to Nihon no shihon shijo (Main banks and Japanese capital markets), *Zenginkyo Kinyu*, August, 11-19.

Miwa, Y. (1996). *Firms and Industrial Organization in Japan*, London: MacMillan Press.

Modigliani, F., and Miller, M. (1958). The Cost of Capital, Corporate Finance and the Theory of Investment, *American Economic Review*, 48, 261-297.

Moerke, A. (1999). Does Governance Matter? Performance and Corporate Governance Structures of Japanese keiretsu Groups, in Albach, H., Görtzen, U., and Zobel, R. (eds.), *Information Processing as a Competitive Advantage of Japanese Firms*, Berlin, Edition Sigma, 211-238.

Montalvo, J. and Yafeh, Y. (1994). A micro-economic analysis of technology transfer: The case of licensing agreements of Japanese firms, *International Journal of Industrial Organization*, 12, 227-244.

Morikawa, H. (1992). *ZAIBATSU. The Rise and Fall of Family Enterprise Groups in Japan.* Tokyo, University of Tokyo Press.

Nakatani, I. (1983). Kigyô shudan no keizaiteki imi to ginko no yakuwari [The economic significance of the enterprise groups and the role of banks], *Kin'yû Keizai*, 202, 51-75.

Nakatani, I. (1984). The Economic Role of Financial Corporate Groupings, in M. Aoki (Ed.), *The Economic Analysis of the Japanese Firm*, North-Holland: Elsevier Publishing, 227-258.

Nerlove, M. (1971). A note on error component models. *Econometrica,* 39, 383-396.

Nihon Keizai Kikakuchô (Economic Planning Agency). *Kokumin keizai keisan nenpô* (Annual Report on National Accounts), Tokyo, EPA, several issues.

Nihon Keizai Shinbun Sha (1987). *Kôshasai hakko shijô* (Bond Primary Markets), Tôkyô, Nihon Keizai Shinbun Sha.

Ogishima, Seiji (1994). Recent Trends in Japanese Stock Cross-Holding, Nomura Research Institute, *NRI Quarterly*, Autumn.

Okuda, M. and James, S. (1990). The Japanese bond market, in Growland, D. (ed.), *International Bond Markets,* New York, NY, Routledge.

Otto, S. (1997). Corporate Control-Mechanismen und Stakeholder Relations in Japan im Wandel, *ZfB*, Ergänzungsheft 4, 43-74.

Pascale, R. T. and Rohlen, T. P. (1983). The Mazda Turnaround, *Journal of Japanese Studies,* Vol.9, 219-263.

Pellens, B. (1989). *Der Informationswert von Konzernabschlüssen*, Wiesbaden, Gabler.

Petersen M. A. and Rajan, R.G. (1994). The Benefits of Lending Relationships: Evidence from Small Business Data, *Journal of Finance,* Vol. XLIX, No. 1, 3-37.

Petersen, M. A. and Rajan, R. G. (1995). The Effect of Credit Market Competition on Lending Relationships, *Quarterly Journal of Economics*, Vol. 110, 405-443.

Rajan, R. G. (1992). Insiders and Outsiders: The Choice between Informed and Arm's-Length Debt, *Journal of Finance*, 45, 1367-1400.

Ramakrishnan, S. and Thakor, A. V. (1984). Information reliability and a theory of financial intermediation, *Review of Economic Studies*, 51, 415-432.

Ramseyer, J. (1994). Explicit reasons for implicit contracts: The legal logic to the Japanese main bank system, in M. Aoki and H. Patrick (eds.), *The Japanese Main Bank System*, Oxford, Oxford University Press.

Rosovsky, H. (1961). *Capital Formation in Japan, 1868-1940*, New York, N.Y., Free Press of Glencoe.

Samolyk, K. (1994). Banking Conditions and Regional Economic Performance. Evidence of a Regional Credit Channel, *Journal of Monetary Economics*, 34, 259-278.

Schaede, U. (1990). *Der neue japanische Kapitalmarkt: Finanzfutures in Japan*, Wiesbaden, Gabler.

Schaede, U. (1996). *The 1995 Financial Crisis in Japan*, Japan Information Access Project, papers on Japanese Deregulation, http://www.nmjc.org/jiap/frame.html.

Scher, M. (1997). *Japanese interfirm networks and their main banks*, Basingstoke, MacMillan.

Seger, M. (1997). *Banken, Erfolg und Finanzierung. Eine Analyse für deutsche Industrieunternehmen*, Wiesbaden, Deutscher Universitäts-Verlag, Gabler Verlag.

Semkow, B. (1992). Japan's 1992 'big bang' and other financial reforms, *Journal of Economic Behavior and Organization*, 11, 399-422.

Sharpe, S. (1990). Asymmetric Information, Bank Lending and Implicit Contracts: A Stylized Model of Customer Relationships, *Journal of Finance*, 45, 1069-1087.

Sharpe, S. (1995). Banking Capitalization, Regulation and the Credit Crunch: A Critical Review of the Research Findings, Federal Reserve Board, *Finance and Economics Discussion Series*, 95/20.

Sheard, P. (1985). *Main Banks and Structural Adjustment*, Australia-Japan Research Center, Research Paper 129.

Sheard, P. (1989). The Main Bank System and Corporate Monitoring and Control in Japan, *Journal of Economic Behavior and Organization*, Vol. 11, 399-422.

Sheard, P. (1994). Bank Executives on Japanese Corporate Boards, *BOJ Monetary and Economic Studies*, Vol. 12, No. 2, 85-121.

Shimizu, Y. (1997a). Speculative Bubbles, Depression and the Monetary Policy in Japan. *Hitotsubashi Journal of Commerce and Management*, 32, 23-58.

Shimizu, Y. (1997b). Nippon no Kinyû to Shijô Mechanism, Tôyôkeizai Shinposha.

Shockley, R. and Thakor, A. V. (1993). *Information content of commitments to lending in the future: Theory and evidence on the gains from relationship banking*, Working paper, Bloomington, Indiana, Indiana University, January.

Shôji Hômu Kenkyûkai (1989). *Kansayaku Seido no Unyô: Seido Unyô ni kansuru Ankêto Chôsa no Bunseki (Operation of the Statutory Auditor System: Analysis of a Questionnaire Study regarding Operation of the System)*, Tôkyô, Shôji Hômu Kenkyûkai.

Steinbrenner, J. O. (1997). Japanische Unternehmensgruppen: Organisation, Koordination und Kooperation der Keiretsu. Stuttgart, Schäffer-Poeschel.

Stiglitz, J. and Weiss, A. (1981). Credit rationing with imperfect information. *American Economic Review*,71, 393-410.

Stiglitz, J.(1985). Credit markets and the control of capital, *Journal of Money, Credit, and Banking*, 17, 133-152.

Stiglitz, J., and Weiss, A. (1981). Credit rationing with imperfect information, *American Economic Review*, 71, 393-410.

Suzuki, Y. (1987). *The Japanese Financial System*, Oxford, Clarendon Press.

Teranishi, J. (1993). Emergence and establishment of the financial system in postwar Japan: Government intervention, indirect financing and the corporate monitoring system, Paper prepared for the World Bank project, *Strategies for Rapid Growth: Public Policy and the Asian Miracle.*

Teranishi, J., Goto, A., and Serizawa, K. (1975). Shikin Shijo to Kigyô Gurûpû. (Financial Markets and Corporate Groups), *Keizai Hyoron*, November 1975.

The Economist (1998). *The Japan Puzzle*, The Economist, 21.03.1998, 15 and 25-27.

Tobin, J. (1969). A General Equilibrium Approach to Monetary Theory, *Journal of Money Credit and Banking*, 1, 15-29.

Tôkyô Bengôshikai Kaishahôbu (1986). *Torishimariyakkai Ketsugi Jikô no Handan Kijun (Standards of Judgement for Decision Items at Directors Meetings),* Tôkyô, Shôji Hômu Kenkyûkai.

Townsend, R. (1979). Optimal Contracts and Competitive Markets with Costly State Verification, *Journal of Economic Theory*, 21, 265-293.

Tröge, M. (1999). *Monitored Finance, Usury and Credit Rationing*, mimeo Berlin Science Center.

Ueda, K. (1989). Wagakuni no kabuka suijun ni tsuite (On the Price Level of Japanese Stocks), *Nihon Keizai Kenkyu*,18, March.

Vihriälä, V. (1996). *Theoretical Aspects to the Finnish Credit Cycle*, Bank of Finland Discussion Papers 8/96.

Wakita, Y. (1983). Wagakuni no Kashidashi Shijô to Keiyaku Torihiki (The Japanese Credit Market and Contract Theory), *Kinyû Kenkyû,* Bank of Japan, March.

Wallace, T.D., and Hussain, A. (1969). The use of error component models in combining cross-section and time-series data, *Econometrica*, 37, 55-72.

Wallich, H. C. and Wallich, M. I. (1976). Banking and Finance, in Patrick, H. and Rosovsky, H. (eds.) *Asia's New Giant,* Washington DC, Brookings Institute.

Wansley, J. W.; Elayan, F. A. and Collins, M. C. (1992). *Investment Opportunities and firm quality: An empirical investigation of the information in bank lines of credit*, Working paper, University of Tennessee, May.

Weinstein, D. and Yafeh, Y. (1998). On the Cost of a Bank-Centered Financial System: Evidence from the Changing Main Bank Relations in Japan. *The Journal of Finance,* Vol. LIII, No. 2, 635–672.

Whited, T. (1992). Debt, Liquidity Constraints, and Corporate Investment: Evidence From Panel Data, *Journal of Finance*, 47, 1425-1460.

Williamson, S. (1987). Financial Intermediation, Business Failures, and the Real Business Cycles. *Journal of Political Economy,* 95, 1196-1216.

Yabushita, S. (1992). Main Bank to jôhô no riron (Main banks and information theory), in Horiuchi, A. and Yoshino, N. (eds.), *Gendai nihon no kinyû bunseki (Structural analysis of today's Japanese financial system)*, Tôkyô, Tôkyô University Press, 9-38.

Yabushita, S. and Inoue, A. (1993). The Stability of the Japanese Banking System: A Historical Perspective, Journal of the Japanese and International Economies, 7, 387-407.

Yafeh, Y. (1995). Corporate ownership, profitability and bank-firm ties: Evidence from the American occupation reforms in Japan, *Journal of the Japanese and International Economies*, 9, 154-173.

Yafeh, Y. and Yosha, O. (1999). *Large Shareholders and Banks: Who Monitors and How*, Working Paper, The Hebrew University of Jerusalem.

Yang, J. (2000). *Aspekte eines Investitionsmodells innerhalb der q-Theorie unter Berücksichtigung von Bankbeziehungen deutscher Unternehmen und Risikoanalyse*, Wiesbaden, Gabler-Verlag.

Yamaura, H. (1990). EC shuyô 2-koku no kaisha kikô no hikaku bunseki to waga-kuni kansayaku seido no mondaiten (Comparative analysis of the corporate structure of the two most important countries in the EC and problems of the Japanese kansayaku system). *Keiei kôdô*, Vol. 5, No. 4, 39-47.

Yamaura, H. (1994). Waga-kuni no kaisha naibu monitaringu kikô no kokusai-teki ichi-zuke to kansa seido no mondaiten (Internal monitoring structures of Japanese corporations in the light of an international comparison and problems of the audit system). *Kokumin keizai zasshi*, Vol. 145, No. 4, 16-28.

Yatsunami, A. (1999). *Financial Big Bang in Japan*, Distinguished Lecturer Series Paper, WHU Center for Japanese Studies, Vallendar, forthcoming.

Zenginkyô (1994). *The Banking System in Japan*, Tôkyô, IBI Inc.

Contributions to Economics

Peter Michaelis/Frank Stähler (Eds.)
Recent Policy Issues in Environmental and Resource Economics
1998. ISBN 3-7908-1137-8

Jessica de Wolff
The Political Economy of Fiscal Decisions
1998. ISBN 3-7908-1130-0

Georg Bol/Gholamreza Nakhaeizadeh/ Karl-Heinz Vollmer (Eds.)
Risk Measurements, Econometrics and Neural Networks
1998. ISBN 3-7908-1152-1

Joachim Winter
Investment and Exit Decisions at the Plant Level
1998. ISBN 3-7908-1154-8

Bernd Meyer
Intertemporal Asset Pricing
1999. ISBN 3-7908-1159-9

Uwe Walz
Dynamics of Regional Integration
1999. ISBN 3-7908-1185-8

Michael Carlberg
European Monetary Union
1999. ISBN 3-7908-1191-2

Giovanni Galizzi/Luciano Venturini (Eds.)
Vertical Relationships and Coordination in the Food System
1999. ISBN 3-7908-1192-0

Gustav A. Horn/Wolfgang Scheremet/ Rudolf Zwiener
Wages and the Euro
1999. ISBN 3-7908-1199-8

Dirk Willer
The Development of Equity Capital Markets in Transition Economies
1999. ISBN 3-7908-1198-X

Karl Matthias Weber
Innovation Diffusion and Political Control of Energy Technologies
1999. ISBN 3-7908-1205-6

Heike Link et al.
The Costs of Road Infrastructure and Congestion in Europe
1999. ISBN 3-7908-1201-3

Simon Duindam
Military Conscription
1999. ISBN 3-7908-1203-X

Bruno Jeitziner
Political Economy of the Swiss National Bank
1999. ISBN 3-7908-1209-9

Irene Ring et al. (Eds.)
Regional Sustainability
1999. ISBN 3-7908-1233-1

Katharina Müller/Andreas Ryll/ Hans-Jürgen Wagener (Eds.)
Transformation of Social Security: Pensions in Central-Eastern Europ
1999. ISBN 3-7908-1210-2

Stefan Traub
Framing Effects in Taxation
1999. ISBN 3-7908-1240-4

Pablo Coto-Millán
Utility and Production
1999. 3-7908-1153-X

Frank Riedel
Imperfect Information and Investor Heterogeneity in the Bond Market
2000. ISBN 3-7908-1247-1

Kirsten Ralf
Business Cycles
2000. ISBN 3-7908-1245-5

Michele Bagella/Leonardo Becchetti (Eds.)
The Competitive Advantage of Industrial Districts
2000. ISBN 3-7908-1254-4

Frank Bohn
Monetary Union and Fiscal Stabilit
2000. ISBN 3-7908-1266-8

Michael Malakellis
Integrated Macro-Micro-Modelling Under Rational Expectations
2000. ISBN 3-7908-1266-8

Stefan Baumgärtner
Ambivalent Joint Production and the Natural Environment
2000. ISBN 3-7908-1290-0

Henri Capron, Wim Meeusen (Eds.)
The National Innovation System of Belgium
2000. ISBN 3-7908-1308-7